Game Theory

and

Experimental Games

The Study of Strategic Interaction

by

ANDREW M. COLMAN

PERGAMON PRESS

OXFORD · NEW YORK · TORONTO · SYDNEY · PARIS · FRANKFURT

U.K.	Pergamon Press Ltd., Headington Hill Hall, Oxford OX3 0BW, England
U.S.A	Pergamon Press Inc., Maxwell House, Fairview Park, Elmsford, New York 10523, U.S.A.
CANADA	Pergamon Press Canada Ltd., Suite 104, 150 Consumers Road, Willowdale, Ontario M2J 1P9, Canada
AUSTRALIA	Pergamon Press (Aust.) Pty. Ltd., P.O. Box 544, Potts Point, N.S.W. 2011, Australia
FRANCE	Pergamon Press SARL, 24 rue des Ecoles, 75240 Paris, Cedex 05, France
FEDERAL REPUBLIC OF GERMANY	Pergamon Press GmbH, 6242 Kronberg-Taunus, Hammerweg 6, Federal Republic of Germany

First edition 1982

Library of Congress Cataloging in Publication Data
Colman, Andrew M.
Game theory and experimental games.
(International series in experimental social psychology; v. 4)
Bibliography: p.
1. Game theory.   I. Title.   II. Series.
QA269.C59   1982   519.3   82–461

British Library Cataloguing in Publication Data
Colman, Andrew M.
Game theory and experimental games.—(International series in experimental social psychology; v. 4)
1. Game theory
I. Title   II. Series
519.3   QA269

ISBN 0–08–026070–5 (Hardcover)
ISBN 0–08–026069–1 (Flexicover)

Printed in Great Britain by The Anchor Press Ltd., Essex

Introduction to the Series

Michael Argyle

Social psychology is in a very interesting period, and one of rapid develop-ment. It has survived a number of "crises", there is increased concern with external validity and relevance to the real world, the repertoire of research methods and statistical procedures has been greatly extended, and a number of exciting new ideas and approaches are being tried out.

The books in this series present some of these new developments; each volume contains a balance of new material and a critical review of the relevant literature. The new material consists of empirical research, research procedures, theoretical formulations, or a combination of these. Authors have been asked to review and evaluate the often very extensive past literature, and to explain their new findings, methods or theories clearly.

The authors are from all over the world, and have been very carefully chosen, mainly on the basis of their previous published work, showing the importance and originality of their contribution, and their ability to present it clearly. Some of these books report a programme of research by one individual or a team, some are based on doctoral theses, others on conferences.

Social psychologists have moved into an increasing number of applied fields, and a growing number of practitioners have made use of our work. All the books in this series will have some practical application, some will be on topics of wide popular interest, as well as adding to scientific knowledge. The books in the series are designed for advanced undergraduates, graduate students and relevant practitioners, and in some cases for a rather broader public.

We do not know how social psychology will develop, and it takes quite a variety of forms already. However, it is a great pleasure to be associated with books by some of those social psychologists who are developing the subject in such interesting ways.

Preface

THE primary aim of this book is to provide a critical survey of the essential ideas of game theory and the findings of experimental research on strategic interaction. In addition, I have reported some new experiments using lifelike simulations of familiar kinds of strategic interactions, and included discussions of recent applications of game theory to the study of voting, the theory of evolution, and moral philosophy. The time has (alas) long since passed when a single person could reasonably hope to be an expert on all branches of game theory or on all of its applications, and I have not achieved the impossible. But I thought it worthwhile, nonetheless, to aim for a fairly comprehensive coverage of important topics, with particular emphasis on those which seem to be most relevant to naturally occurring strategic interactions.

Game theory and the experimental gaming tradition have grown up in relative isolation from each other. Game theorists, in general, remain largely oblivious of the empirical studies that have been inspired by the theory, and experimental investigators have tended to assume that the nuts and bolts of the theory do not concern them. Both parties are the losers from this divorce, and I have therefore tried to contribute towards a reconciliation by examining in detail, for the first time in a single volume, both sides of the story.

My goal has been to introduce and evaluate the fundamental theoretical ideas, empirical findings, and applications as clearly as possible without over-simplifying or side-stepping important difficulties. In so far as I have succeeded, this is the kind of book I should have liked to have read when I first became interested in game theory and experimental games, or better still, before I had developed any interest in these matters. Wherever possible, I have attributed seminal contributions to their originators and cited the original sources: ideas are almost invariably expressed more clearly and forcefully by their inventors or discoverers than by subsequent commentators. But I have also cited many useful review articles which will be of assistance to readers wishing to pursue particular topics in depth.

The most important chapters for social psychologists and others whose primary interest is in such strategic phenomena as cooperation, competition, collective equilibria, self-defeating effects of individual rationality, coalition formation, threats, altruism, spite, escalation, social entrapment, and so forth, are Chapters 1, 2, 3, 6, 7, 8, and 9. Mathematically inclined readers should pay special attention to Chapters 4, 8, 10, and 11, and to Appendix A in which the minimax theorem is rigorously proved. The chapters which are most relevant to sociology, economics, and politics are Chapters 1, 2, 6, 8, 10, and 11. Biological applications are discussed in Chapter 12, but Chapters 1, 2, 6, and 8 provide a necessary background to it. Philosophical applications are dealt with primarily in Chapter 13, to which Chapters 1, 2, 6, 8, and 12 provide the necessary background. Most of the translations from original French and German sources in Chapter 13 and elsewhere are my own.

I am indebted to a number of people who contributed to this book in various indirect ways. In particular, I am grateful to Michael Argyle, Alan Baker, Barbara Barrett, Dorothy Brydges, Roy Davies, Julia Gibbs, Gabriele Griffin, John Lazarus, Nicholas Measor, Richard Niemi, Ian Pountney, Albert W. Tucker, Diane Williams, Bill Williamson, and the Research Board of the University of Leicester. I should be delighted to receive comments from readers, indicating their reactions to the final product.

A.M.C.

Contents

Background

1

Introduction

1.1 Intuitive Background

GAME theory is concerned with the logic of decision making in social situations in which the outcomes depend upon the decisions of two or more autonomous agents. An essential feature of such situations is that each decision maker has only partial control over the outcomes. It is immediately obvious from this informal definition that games like chess and poker fall within the ambit of game theory, but that many other activities commonly referred to as games, such as patience (solitaire), space invaders, doll-play, make-believe tea parties, and model building do not. More importantly, there are numerous social situations which are never referred to as games in everyday speech but are nonetheless games in the terminology of game theory. Political, military, economic, and other social conflicts are often regarded as games in the technical sense, together with social situations lacking strict conflicts of interests. A phrase such as "the theory of interdependent decision making" captures the essence of the theory in a less misleading way, but it is too late to rename game theory without risking even greater confusion.

Traditional theories in social psychology and related fields lack the necessary concepts with which to deal rigorously with interdependent decision making. They have tended to follow the theoretical models of classical physics in which the behaviour of inanimate objects is explained in terms of responses to external forces. But inanimate objects do not make deliberate decisions, and their behaviour is not governed by any assumptions about how other objects are likely to behave. One-way causal models may be appropriate for explaining certain involuntary human actions, such as kicking the air in response to a tap on the patellar tendon or winking an eye in response to the invasion of a speck of dust. But a person may kick the air voluntarily as an act of symbolic aggression in the hope of gaining some strategic advantage during a competitive interaction, and another may deliberately wink an eye in an attempt to increase the intimacy of a personal relationship. These are deliberate decisions the outcomes of which depend

3

upon the decisions of other people as well, and one-way causal models are unlikely to come to grips with them. Most psychologically interesting forms of social behaviour result from deliberate decisions, and some are amenable to game theory analysis.

A game theory model can be constructed only if the options available to each decision maker and their possible consequences are well defined, and each decision maker has consistent preferences among the potential outcomes. The best way of gaining an intuitive understanding of the subject matter of game theory is by considering some typical examples.

Head On

Two people are running in opposite directions along a corridor. They are set on a collision course, and each would prefer to avoid a collision. Each may be assumed to have only three options from which to choose: swerve to the left, swerve to the right, or keep going straight ahead, and they have to make their decisions immediately. If both keep going straight ahead, or if both swerve towards the same side of the corridor, they will collide; all other combinations of decisions lead to non-collision outcomes. The outcomes obviously depend upon the decisions of both runners, whose interests coincide exactly.

Price War

Three retail companies are each trying to corner a larger slice of some specified market for which they compete. Each has to decide, in ignorance of the decisions of the others, whether to reduce the price of its product or to maintain it at its present level. If all three companies reduce prices, increased sales will exactly offset reduced profits per unit sale and none will gain or lose anything. The *status quo* will also be preserved if all three decide to maintain prices. Other combinations of decisions result in gains for one or two of the companies at the expense of the other(s). The interests of the companies are diametrically opposed to one another.

Angelo and Isabella

This poignant example is taken from Shakespeare's *Measure for Measure* (II.iv): Angelo is holding a prisoner whom he intends to execute. The prisoner's sister, Isabella, arrives to plead for her brother's life. Angelo is sexually attracted to her, and at length he proposes the following ungentlemanly bargain: "You must lay down the treasures of your body" in order to save your brother. Isabella initially declines the offer—"More than our brother is our chastity"—whereupon Angelo complicates the game by threatening not merely to kill the prisoner as he had originally intended, but to subject him to a lingering death unless she submits. At this point Isabella has a choice between submitting and refusing, and whichever option she chooses there are three courses open to Angelo: to spare the prisoner's life, to execute him humanely, or to subject him to a lingering death. All but one of the possible pairs of decisions (Isabella's refusal coupled with a reprieve from Angelo) lead to unattractive outcomes from Isabella's point of view, although some are clearly worse than others. But both protagonists prefer some possible outcomes to some others since both would rather see the prisoner executed humanely than cruelly, other things being equal. Their interests are partly opposed and partly coincident. (See Schelling, 1960, p. 140.)

Game theory deals with situations such as these by abstracting their formal, logical properties. Its major goal is to find solutions to games. A solution is a specification of the decisions which will be made and the outcome which will therefore be reached if the decision makers are rational according to criteria laid down by the theory. This is the *normative* or *prescriptive* goal of the theory. In some classes of games, logically compelling formal solutions can be found, and these are often far from obvious. In others, formal solutions are elusive, but the games can be analysed by informal methods. Even in the absence of satisfactory formal solutions, a considerable insight into interdependent decisions can often be attained. Certain important features of individual and collective rationality, cooperation and competition, trust and suspicion, threats and commitments cannot even be clearly formulated without the concepts of game theory. The conditions under which decision makers behave in accordance with the normative prescriptions of the theory or deviate from them is a problem for empirical research; the theory is not primarily *positive* or *descriptive*.

1.2 Abstract Models: Basic Terminology

A model of a specific "slice of life" in game theory terms is a purely imaginary object. The social reality, by its very nature, bristles with complexity and can be grasped only vaguely; it is therefore replaced by an idealized and deliberately simplified formal structure which is amenable to purely logical analysis. The conclusions of such an analysis apply not to the social reality itself but to an abstraction based upon certain properties which it is thought to possess. These conclusions are true provided only that they are logically consistent; their truth has nothing to do with their correspondence to the original social reality. If the abstract model does not correspond with reality in important details, however, it is of little practical value, and this is also the case if it does not yield insights that transcend our common-sense understanding of the social reality which it is intended to represent.

Isaac Newton postulated unreal objects which have mass but no size and are attracted to one another according to a simple formula. Game theorists postulate unreal decision makers who are rational and error-free and choose options according to specified principles. The power of either theory to account for the behaviour of real objects or decision makers under various conditions can be tested through experiments.

The autonomous decision makers in a game are called *players*. From the examples in the previous section it is clear that the players may correspond to individual human beings or corporate decision-making bodies, and in some applications of the theory the players represent non-human animals. The essential quality of any player is the ability to choose, in other words to make decisions. These decisions or choices are sometimes called *moves*.

In some games, the moves made by ordinary decision makers do not lead to definite outcomes; the outcomes may partly depend upon the invisible hand of chance. In order to handle games like these, a fictitious player called Nature is invoked, and it is usually assumed that Nature moves disinterestedly according to the laws of probability. Poker is a typical example: provided that the cards are properly shuffled, Nature makes the first move by arranging the deck in a particular order, each possible arrangement occurring with equal probability. In many non-recreational games, Nature also plays a part.

A game must involve at least two players, one of whom may be Nature in the widest interpretation of game theory, otherwise there can be no interdependence of choice, and the total number of players is assumed to be finite except in esoteric applications of the theory. Each player must, in addition, have at least two options from which to choose; an agent with only one way of acting has no effect on the outcome of the game and can safely be ignored. In defining a player's options it is important to realize that doing nothing is sometimes a definite move, provided that there is at least one other option available.

The rules of the game determine what options are available to each player at each move, and the *outcomes* associated with any complete set of moves by all the players. The rules of chess, for example, define the permissible moves and the three possible outcomes: White wins, Black wins, or Draw. The three hypothetical games outlined in the previous section have reasonably obvious rules. The outcome of Head On is either a collision or the avoidance of a collision, and in Price War the outcomes are various distributions of profits and losses among the three companies. Angelo and Isabella is a slightly more complicated case. One possible outcome is this: Isabella surrenders her chastity to Angelo, and her brother is subjected to a lingering death.

This is, from Isabella's point of view, the worst of all possible outcomes of the game. It is assumed in general that the players have consistent preferences among the possible outcomes. At the very least, each player must be capable, in principle, of arranging the possible outcomes in order of preference from "worst" to "best" with equal ranks alloted to those between which the player is indifferent. It is then possible to assign numerical *payoffs* to the outcomes, indicating each player's relative preferences among them. In Angelo and Isabella, for example, it is possible to assign Isabella's lowest payoff, say 1, to her least-preferred outcome, 2 to her next-to-worst outcome—presumably a surrender of her chastity to Angelo coupled with a humane execution of her brother—and so on up to her favourite outcome, namely a preservation of her chastity and a reprieve for her brother. Angelo's lowest payoff may (reasonably) be assigned to the outcome in which Isabella withholds her sexual favours and her brother is reprieved, and his highest to the one in which Isabella submits and her brother is humanely executed as originally planned. Ordinal payoff scales like these are sufficient for solving some games, but in others the payoffs must be measured on interval scales. In other words, it is sometimes necessary to know not merely which outcomes are preferred by each player to which others, but also how intense the preferences are compared with one another. In either event, a *payoff function* is defined for each player. A payoff function is nothing more than a set of (ordinal or interval) payoffs linked to the possible outcomes of the game; it shows what payoff the player will receive in every conceivable contingency.

A *pure strategy* is a complete plan of action, specifying in advance what moves a particular player will make in all possible situations that might arise during the course of the game. If each player has only one move to make, and if it is made in ignorance of the moves chosen by the other players, then the players' pure strategies are simply the options from among which they have to choose. In Head On, for example, each player has three pure strategies: swerve left, swerve right, and continue straight ahead. In Price War, each player's pure strategies are simply to reduce prices and to maintain them. In other cases the players' pure strategies are more complex than their options. In Angelo and Isabella, Isabella's pure strategies are simply: submit (*S*) and don't submit (*DS*). But Angelo's move is made after—and in full knowledge of—hers, and his pure strategies can be specified only by stating in advance what move he will make in response to each of her possible moves. Thus, although Angelo has only three options according to the rules of the game, to reprieve (*R*), execute humanely (*E*), and torture Isabella's brother to death (*T*), he has nine pure strategies as follows:

(1) If *S*, choose *R*; if *DS*, choose *R*.
(2) If *S*, choose *R*; if *DS*, choose *E*.
(3) If *S*, choose *R*; if *DS*, choose *T*.
(4) If *S*, choose *E*; if *DS*, choose *R*.

(5) If S, choose E; if DS, choose E.
(6) If S, choose E; if DS, choose T.
(7) If S, choose T; if DS, choose R.
(8) If S, choose T; if DS, choose E.
(9) If S, choose T; if DS, choose T.

Angelo's promise (or threat, depending on how it is viewed) is to choose his third pure strategy, but he is not bound by the rules of the game to honour it if his bluff is called. In Shakespeare's original play, Angelo's untrustworthiness becomes clear as the story unfolds and it transpires that he has in fact chosen his fifth pure strategy in spite of his promise.

If each player in a game chooses a pure strategy, then no one actually relinquishes any freedom of action. All of the moves available to the players in the original *extensive form* of the game are preserved in this *normal form*. Thus, a game involving a sequence of moves can be represented statically, as though the outcome were determined by a single choice on the part of each player, without affecting the strategic possibilities inherent in the situation. The pure strategies chosen by the players determine a unique sequence of moves and a definite outcome. On the basis of the pure strategies, a referee could in principle make all the moves in accordance with the players' plans and discover the outcome. But this is a practical impossibility in games of even moderate complexity. In chess, for example, some 160,000 distinguishable positions can arise after only two moves by each player, and all the paper in the world would not suffice to write down a player's pure strategy for an entire game. In complicated games, the extensive form of the model more closely resembles the way the players actually think; the normal form, though logically equivalent, may be psychologically unreal (Argyle, 1969, p. 199).

Two technical terms, which need to be carefully distinguished, are used to describe the players' state of knowledge during the game. In a game of *complete information*, the players know their own permissible moves or pure strategies and payoff functions as well as those of the other players. In other words, they know not only the rules of the game, which prescribe the possible choices and outcomes, but also the preference scales of the other players. There is also a tacit assumption that each player knows that the others have complete information. The assumption of complete information is usually made, but models involving incomplete information or misperception are sometimes used. In these games, each player is usually assumed to know the rules of the game but not the preference scales of the other players. A game of *perfect information* has this property: the players move one at a time and in full knowledge of any moves that have been made previously. Chess, and Angelo and Isabella, are games of perfect information, but Head On and Price War are obviously not.

1.3   Skill, Chance, and Strategy

It is possible to classify games according to certain family resemblances. This is useful because the method of analysing a specific game can often be applied without modification of the fundamental ideas to other games which belong to the same general class; only a relatively few games need therefore to be studied in order to understand a much larger number. Perhaps the most basic criterion of classification concerns the factors which affect the outcomes of the game. We may accordingly distinguish between games of skill, games of chance, and games of strategy.

Games of skill and chance are classes of one-person games, while games of strategy involve two or more decision makers. Games of skill are conventionally referred to in the literature as "individual decision making under certainty", and games of chance are called "individual decision making under risk or uncertainty" or "one-person games against Nature", but such a natural distinction deserves less cumbersome terminology.

The defining property of a game of skill is the existence of a single player who has complete control over the outcomes: each of the player's possible actions leads to a single certain outcome. Responding to an IQ test, for example, can be modelled by a game of skill. Since choices of other decision makers and Nature have no effects on the outcomes of these games, they constitute a degenerate class from which the element of interdependence of choice is lacking, and they do not, strictly speaking, qualify as games in the technical sense outlined in Section 1.2. For the sake of completeness, however, and also to place other one-person games in their proper theoretical perspective, a brief discussion of decisions involving certainty is included in Chapter 2.

Games of chance are models of one-person decisions whose outcomes are influenced by Nature. In these games the decision maker does not control the outcomes completely, and his or her choices do not therefore lead to outcomes that can be predicted with certainty. The outcomes of a game of chance depend partly on the choices of the decision maker—the person to whom the adjective "one-person" refers—and partly on those of the fictitious player, Nature. Although they are one-*person* games, they are modelled in terms of two *players*, and they can be considered to fall within the domain of game theory.

Depending on how Nature's moves can be interpreted, games of chance are further subdivided into those involving risk and those involving uncertainty. In a situation requiring risky decisions, the decision maker knows the probabilities of each of Nature's moves, and therefore of the outcomes associated with each choice which he or she might make. Russian roulette is a clear example of a game of chance involving risk: for any number of live bullets between one and five which the decision maker chooses to place in the revolver, the probability of each possible outcome—life or death—is known, although it is never entirely certainly certain what will happen. The theory of games of chance is, in essence, the theory of gambling. All games of this type can be satisfactorily solved, as will be shown in Chapter 2, provided that the player's preferences among the outcomes are not inconsistent.

There are other games of chance, however, in which probabilities cannot meaningfully be assigned to Nature's moves, and these are modelled in terms of decisions involving uncertainty. A doctor who needs to choose a method of treatment for a patient with an extremely rare complaint may, for example, have no basis upon which to judge the relevant probabilities. Nature has a hand in this game because the outcomes of the doctor's decisions are not certain, but in this case even the probabilities associated with Nature's moves are unknown. Ingenious solutions have been devised for games of this type, but it will be argued in Chapter 2 that the principles on which they are based are not entirely satisfactory.

Games of strategy are concerned with situations in which the outcomes depend upon the choices of two or more decision makers, each of whom has partial control over the outcomes. Since the players cannot meaningfully assign probabilities to one another's moves, the decisions are made under uncertainty rather than risk, although Nature also plays a part in some games of strategy by introducing an element of risk. The bulk of the literature on game theory is concerned with games of strategy, and it is fair to say that the theory came into its own only with the discovery of methods for solving these games.

Games of strategy have particular relevance to understanding social interaction, and most of this book is devoted to the theory and research associated with them. They can be subdivided into two-person and multi-person games according to the number of players (excluding Nature) involved. Both two-person and multi-person games can be further classified in terms of certain structural properties. One important structural property is the way in which the players' payoff functions are related to one another. There may be complete coincidence between the players' interests as in Head On, strict opposition of interests as in Price War, or partial coincidence and partial opposition of interests as in Angelo and Isabella. Those in which the players' interests coincide are called *coordination games*, and research on situations of this type is discussed in Chapter 3. Strictly competitive games are known technically as *zero-sum* games, and Chapters 4 and 5 are devoted to them. Most recreational games are zero-sum since if one player wins, the other must lose, and the payoffs therefore add up to zero in each outcome, but relatively few social encounters in everyday life have this structural property. The rest of this book is concerned mainly with two-person and multi-person games in which the interests of the players are neither strictly coincident nor strictly opposed. These are known technically as *mixed-motive* games. Two-person, zero-sum games can be convincingly solved, provided that each player has only a finite number of moves, and empirical research has provided some information about the way people behave in such situations. Other categories of games of strategy are not always completely soluble in a formal sense, but they provide insights into important areas of interdependent choice and have attracted a great deal of attention from researchers.

1.4   Historical Background

The historical forerunner of the theory of games of strategy was the theory of games of chance. This earlier theory originated in the seventeenth century from attempts to solve practical problems of gambling raised by members of the dissolute French nobility. From these humble origins sprang the theory of probability, on whose foundations the exalted disciplines of statistics, population genetics, and quantum physics were later erected.

The theory of games of chance is traced by most historians to 1654, to a correspondence between Pascal and Fermat concerning the misfortunes of the French nobleman, the Chevalier de Méré. De Méré had written to Pascal explaining how he had won a considerable sum of money over a period of time by betting even odds that at least one six would come up in four throws of a die, only to lose it all by betting that at least two sixes would come up in 24 throws. According to de Méré's faulty calculations, the second bet ought to be just as favourable as the first, but the correct probabilities are in fact 0.5177 and 0.4914 respectively, that is, the first is favourable but the second is not. De Méré must have been a remarkably industrious gambler to have discovered by sheer bitter experience that the second bet is unfavourable, in other words to have distinguished empirically between a probability of 0.4914 and 0.5000. The elementary principles of probability and expected value calculations were developed in order to solve de Méré's and other similar problems. The story is entertainingly told, with extracts from the Pascal-Fermat letters, by David (1962).

The first important contribution to the theory of games of strategy was made by the German mathematician, Zermelo, in 1912. Zermelo managed to prove that every strictly competitive, two-person game of perfect information possesses either a guaranteed winning pure strategy for one of the players, or guaranteed drawing pure strategies for both. This result applies, for example, to zero-sum board games like chess in which each player knows at every move what moves have been made previously, but it provides no method for finding the winning (or drawing) strategies. It is still not known whether White (or conceivably Black) has a winning strategy in chess, or whether both players have drawing strategies, although most grandmasters would guess the latter.

Much of the groundwork of the theory of games of strategy was laid in a series of papers by Borel between 1921 and 1927 (see Fréchet, 1953). But Borel was unable to prove the fundamental minimax theorem which lies at the heart of game theory (see Chapter 4 and Appendix A); in fact he rashly predicted that no such theorem could be proved. The minimax theorem was promptly proved by the German mathematician, John von Neumann, in 1928. In 1934, independently of von Neumann, the British mathematician, R. A. Fisher, who is remembered chiefly for his seminal contributions to experimental design and statistics, proved the minimax theorem for the special case in which each player has only two pure strategies. Game theory did not, however, attract much attention until the publication in the United States of von Neumann and Morgenstern's classic *Theory of Games and Economic Behavior* in 1944. This book stimulated a great deal of interest amongst mathematicians and mathematically sophisticated economists, but it was a later text by Luce and Raiffa (1957), entitled *Games and Decisions*, that made the theory accessible to a wide range of social scientists and psychologists.

Only a handful of empirical investigations of the behaviour of people in strategic interactions appeared before 1957, but the publication of Luce and Raiffa's book was followed by a steady growth of experimental gaming as a field of research. The first comprehensive review of experimental gaming studies was given by Rapoport and Orwant in 1962, by which time 30 experiments had appeared in print. By 1965, experimental gaming had become sufficiently popular for the *Journal of Conflict Resolution* to decide to devote a special section in each issue to such research, and by 1972 more than 1000 empirical studies had been published (Wrightsman, O'Connor, and Baker, 1972; Guyer and Perkel, 1972). Towards the mid-1970s, however, many researchers began to express dissatisfaction with experimental gaming, which was then dominated by extremely abstract versions of one particular two-person game known as the Prisoner's Dilemma (see Chapters 6 and 7). This crisis of confidence in experimental gaming was highlighted by an acrimonious debate in the *European Journal of Social Psychology* initiated by Plon (1974). The volume of experimental gaming research nevertheless continued to increase, with less emphasis on relatively artificial games like the standard Prisoner's Dilemma and more on multi-person games. By the early 1980s some 1500 experimental gaming studies had appeared.

Theoretical work on games of strategy also accelerated during the 1960s and 1970s. In 1971 the *International Journal of Game Theory* was founded. Soviet scholars began to make significant contributions to game theory in the late 1960s and 1970s (see Robinson, 1970; Vorob'ev, 1977). Numerous applications of the theory to economics have been published since the 1950s, but applications to politics, biology, and philosophy, some of which are discussed in the final chapters of this book, are of more recent origin and are attracting increasing interest.

1.5 Summary

Section 1.1 opened with an informal definition of game theory and an outline of the types of social situations to which it applies. An essential property of these situations was shown to be interdependent decision making, and it was argued that game models are more appropriate than the usual one-way causal models for understanding them. Three hypothetical examples were given of social interactions to which game theory might be applied. In Section 1.2 the crucial features of all game models, and their technical names, were outlined with the help of the examples given earlier. Section 1.3 focused on the various categories into which games can be grouped. Games of skill—which constitute a degenerate class, games of chance—which involve either risky gambles or uncertainty, and games of strategy were distinguished. The further subdivision of games of strategy into coordination, zero-sum, and mixed-motive varieties was briefly explained. Section 1.4 was devoted to a synopsis of the historical development of game theory and experimental gaming from the mid-seventeenth century to the present.

2

One-Person Games

2.1 Games Against Nature

THIS chapter contains an introduction to the theory of one-person games. The problems to be discussed are sometimes described by the phrase "one-person games against Nature", and because of their essentially non-social properties, some theorists do not consider them to be "genuine" games. The type of dilemma outlined in the following section, in which Nature has no move affecting the outcomes of the decision maker's choices, is certainly not a game in the strict sense, but it is a limiting case which serves as a useful and logical point of departure. Risky decisions, which are discussed in Sections 2.3 and 2.4, are interdependent in a formal sense although they involve no social interaction, and they are analysed by means of certain key ideas which can be extended to the solution of two-person and multi-person games. The introduction of probability concepts and modern utility theory, in particular, provides an essential background to the solution of more complex games. Since two-person and multi-person games are special categories of decision making under uncertainty, the restricted case of one-person uncertainty discussed in Section 2.5 highlights some simple ideas which reappear in more sophisticated guises in later chapters.

2.2 Certainty

Individual decisions under certainty are games of skill in which the solitary player knows with certainty what the outcome of any strategy or choice will be. The player completely controls the outcomes, which are not affected by any actions on the part of other interested parties or Nature. It is worth pointing out that sports like golf or archery, even if they are regarded as one-person games in which the outcomes are simply the player's possible scores, are not games of pure skill. The reason for this is that the choices which the player makes do not lead to outcomes that are perfectly predictable: there is an element of chance in these games which reflects the player's imperfect control over the outcomes. To put it another way, Nature influences the out-

14

comes to a degree which depends upon the player's level of skill. Solving a crossword puzzle, on the other hand, may be modelled as a game of skill, and so may the tasks of allocating a weekly shopping budget to various categories of goods in the most efficient way and of writing computer programs to perform certain logical operations in the fewest number of steps.

From a theoretical point of view, games of skill are completely devoid of interest: the obvious solution in all cases is to choose the pure strategy that maximizes the payoff, in other words to choose the option that produces the most desirable outcome. But in spite of this, a branch of mathematics known as linear programming, and large tracts of operational research, management science, welfare economics, artificial intelligence, and (non-social) psychology are devoted precisely to these non-games. The explanation is simple: optimal strategies, though theoretically unproblematical, are in practice difficult or impossible to find in complicated cases. The player may wish to maximize the payoff but not know which option to choose in order to do so.

The famous problem of the travelling salesman illustrates the practical difficults that can arise under certainty. Starting from home, the salesman needs to visit ten specified cities before returning to base, and it would be rational to choose the shortest possible route for this round trip. The options, i.e. the various possible routes available tc the salesman, are well defined, and the outcomes in terms of distance can be looked up on a map. Maximizing the payoff is in this case equivalent to minimizing the distance, and the salesman's objective is to choose the best route according to this simple criterion. In the language of linear programming, the solution is found by minimizing the objective function which defines the total distance of the round trip. But there are no fewer than 3,628,800 routes from which to choose! There is a choice of ten cities for the first visit, and for each of these there are nine possibilities for the second visit. There are thus $10 \times 9 = 90$ different ways in which the salesman can begin the round trip by visiting two cities. For each of these 90 "openers", there are eight possibilities for the third visit, and so on. The total number of different round trips which must be considered is therefore $10 \times 9 \times 8 \times 7 \times 6 \times 5 \times 4 \times 3 \times 2 \times 1 = 3,628,800$. The number rises astronomically for cases involving more than ten cities: for eleven it is 39,916,800. It requires more patience than most travelling salesmen possess to examine each alternative and compare its distance with each of the others. But no simple formula has yet been discovered to make this problem trivial from a practical or computational point of view (the necessary mathematics is explained in Franklin, 1980, pp. 151–153), and difficulties of this sort are not untypical in games of skill.

Although games of skill, or decisions under certainty, constitute a vast area of research, nothing more will be said about them here because they involve only one player and are therefore not part of game theory in the strict sense.

2.3 Risk

A risky individual decision is a game of chance in which the solitary decision maker is pitted against the fictitious player, Nature. The decision maker does not know with certainty what moves will be made by Nature, but can assign meaningful probabilities to them, and therefore to the various outcomes of his or her own actions. It was pointed out in Chapter 1 that this is precisely the type of problem confronting a player of Russian roulette.

On the face of it, a decision involving only risk presents no problems. It can apparently be solved by working out, according to elementary principles of probability theory, the *expected value* of each pure strategy. (The rudiments of probability are explained from first principles in, for example, Colman, 1981, chap. 4.) The following imaginary game of chance illustrates the central ideas. A competitor in a television quiz game, after answering several answers correctly, is offered a choice between two alternatives:

(a) A coin will be tossed; if it lands heads, the prize will be £1000, but if it lands tails, no prize will be given.

(b) The competitor may select one of three envelopes, which are known to contain prizes of £900, £300, and £150 respectively, although there is no way of knowing which amounts are in which envelopes.

What is the rational strategy? Intuitively, and also according to the elementary laws of probability, the expected value of (a) is

$$(1/2)(1000) + (1/2)(0) = 500,$$

and the expected value of (b) is

$$(1/3)(900) + (1/3)(300) + (1/3)(150) = 450.$$

The competitor should, according to this criterion, choose (a) because "on average" it pays better than (b), in the sense that if the game were repeated many times, the player would win more money by choosing (a) than (b). It seems obvious to many people that this principle can be generalized to all purely risky decisions.

There are two serious objections to this type of solution, however. In the first place, it is well known that rational human beings do not make risky decisions according to the principle of expected monetary value in all circumstances. Roulette (the French, rather than the Russian variety) is a classic game of chance in which Nature's options and the probabilities with which she chooses them are known, yet millions of people play this game in the full knowledge that the expected value is always negative, otherwise the house would not make a profit. The same applies to pools and raffles, and to one-armed bandits, fruit machines, and other gambling devices found in amusement centres. In all such games, the expected monetary value of not playing at all, which is zero, is higher than that of playing for any stakes. Of course one can cling on to the simple-minded expected value principle by simply defining gambling as a form of irrational behaviour, but in that case what is one to make of insurance? The expected monetary value of insuring one's property is negative for essentially the same reason: insurance companies are out to make a profit, and they calculate their premiums according to estimated probabilities so that their clients are bound to lose "on average". Yet sober citizens who would not succumb to the temptations of fruit machines are seldom invulnerable to those of insurance. The notion of rationality would become unreasonably warped if buying insurance were defined as a form of behaviour that is necessarily irrational.

Secondly, simple examples can be given in which the principle of maximizing expected monetary value appears to break down by generating obviously absurd solutions. The most famous of these is the paradox first presented by the Swiss mathematician Daniel Bernoulli to the St Petersburg Academy in the early eighteenth century. The St Petersburg game is this: a coin is tossed, and if it falls heads, the player is paid one rouble and the game ends. If it falls tails, it is tossed again and this time the player is paid two roubles if the coin is heads. This process is continued for as long as is necessary, doubling the payoff each time, until the player wins something, at which point it stops. Assuming that the house has unlimited funds, what is the expected monetary value of playing this game? Or to put the same question another way, how much should a rational person be willing to pay for the privilege of playing it? According to the "obvious" expected value principle used earlier, the calculation is straightforward. The player wins 1 rouble with a probability of 1/2, 2 roubles with a probability of 1/4, 4 roubles with a probability of 1/8, and so on. The "average" or expected value of playing the game is thus

$$(1/2)(1) + (1/4)(2) + (1/8)(4) + \ldots,$$

and the sum of this endless series is infinite since each of its terms is equal to 1/2. According to the principle, therefore, it would be rational to offer one's entire fortune, such as it may be, in order to play the St Petersburg game once, because "on average" the payoff is worth more than any fortune. But this conclusion is absurd for the following reason: if a player offered anything more than a few roubles for playing the game once, the probability would be high that he or she would simply lose it. According to the expected value principle, this is more than counterbalanced by the very small probability of winning a vastly greater amount through a long series of tails. But the example of the St Petersburg game is fatally damaging to the principle of expected monetary value, as Daniel Bernoulli realized, because it would clearly be irrational to stake one's fortune for the privilege of playing a game when there was a great likelihood of losing it. The question which naturally arises is this: can a better principle be found for solving games involving pure risk?

2.4   Utility Theory

The risky choices of rational decision makers are clearly guided by their preferences among the possible outcomes. Yet, as the examples in Section 2.3 showed, the superficially plausible principle of expected monetary value maximization cannot satisfactorily distinguish between common sense and foolishness in the realm of risky decision making.

Daniel Bernoulli's great contribution was the suggestion, which seems obvious in hindsight, that the values of things to a person are not simply equivalent to their monetary values. A small sum of money may be precious to a pauper but nearly worthless to a millionaire, even if the pauper and the millionaire are one and the same person at different times, for example the day before and the day after winning a massive sweepstake. Decisions are guided by what the outcomes are worth to the decision maker rather than by their simple cash value. Bernoulli (1738) called the subjective desirability of an outcome its "moral worth", and this hypothetical quantity later became known as *utility*.

If the outcomes of a game are merely varying amounts of money, then their utilities must obviously depend upon their monetary values in some lawful manner. Other things being equal, a large amount of money cannot have less utility for a rational person than a smaller amount; any sum of money can be converted into a smaller amount without difficulty by disposing of some of it. But the utility function may not be *linear* with respect to money: an increase of x units of money may not increase a person's utility by the same amount irrespective of the initial capital. Bernoulli suggested a "law of diminishing marginal utility", as it was later called: the greater a person's initial capital, the smaller the utility (or disutility) of any fixed monetary gain (or loss). Bernoulli went further in claiming that the functional relation between utility and monetary value is a simple logarithmic one; this implies that equal increases in utility correspond to equal percentage increases in monetary value, just as logarithms increase by equal steps as their corresponding numbers increase by equal proportions. Historians of psychology (e.g. Boring, 1950) do not usually mention that in this Bernoulli anticipated by more than a century the famous psychophysical law of G. T. Fechner, which propounds the same logarithmic relation between sensation and stimulus intensity in general.

Bernoulli's theory of utility fell into disrepute for two main reasons. The first was its arbitrariness: there seemed to be no logical or empirical reason why the relationship should be logarithmic, or why it should be the same for all people in all circumstances. The second major limitation of Bernoulli's theory was that it provided no way of assigning utilities to non-monetary out-comes. How, for example, can the logarithmic function be used to assign utilities to the outcomes of a game of Russian roulette? The answer is straightforward: it cannot. In order to solve games like this we do, however, need utilities, measured quantitatively on an interval scale. We shall also need such quantitative utilities for solving certain classes of two-person and multi-person games. Recalling the examples used in Chapter 1, however, Bernoulli's theory does not help us to assign interval-scale utilities to pre-serving one's chastity and saving one's brother's life, or even to avoiding a collision in a corridor. But these problems of measurement are not, in princi-ple, insoluble.

Von Neumann and Morgenstern (1944) rehabilitated the notion of utility by proposing a completely different theory which suffered from neither of the major flaws of the old one. The essential ideas are simple and the theory is extremely flexible. Von Neumann and Morgenstern's utility theory is based on the assumption that a player can express a preference (or indiffer-ence) not only between any pair of outcomes, but also between any outcome and any lottery involving a pair of outcomes, or between any two lotteries. To give a concrete interpretation of these ideas, let us consider once again the imaginary television quiz game outlined in Section 2.3. For von Neumann and Morgenstern's utility theory to work, a player must be able to express preferences among the possible outcomes—prizes of £150, £300, £900, and £1000—and also between such pairs of possibilities as this:

(a) £300 for certain; *or*
(b) a 50–50 lottery between £1000 and nothing.

If the player prefers (b) to (a), then a new choice can be offered between a fixed amount of money larger than £300 and (b). It is reasonable to assume that there is *some* amount which any player will consider no less and no more desirable than (b). If this assumption is satisfied, then it is possible to convert the monetary outcomes of the game into utilities, provided only that the player's preferences are consistent. If the amounts involved are not too large, then it is very likely that a person's corresponding utilities will be directly proportional to the monetary values; most people consider £4 to be twice as desirable as £2, and £3 three times as desirable as £1, and this will be reflected in their preferences between lotteries. When large amounts are examined, however, this linear relationship between cash value and utility is likely to break down; £10 million may not seem ten times as desirable as £1 million.

Von Neumann and Morgenstern succeeded in demonstrating two things about these utilities. First, it is always possible in principle to convert a player's consistent preferences among the outcomes of a game into utilities. Secondly, if a player applies the principle of maximization outlined in the previous section to the expected utilities rather than to the expected monetary values of the available strategies, then this player is in fact choosing according to his or her tastes. In contrast to the principle of maximizing expected monetary value, this is called the principle of maximizing *expected utility*.

In the light of the modern theory of utility, it is not necessarily irrational to gamble or to insure one's property. The expected utility of a gamble involving the probable loss of a small stake and the improbable gain of a large prize may be positive even if the expected cash value is negative; and for most people the expected utility of a gamble involving the very likely loss of a small insurance premium and the unlikely repayment of the value of one's home after a fire is positive in spite of its negative expected cash value. It is equally obvious that the paradoxical quality of the St Petersburg game vanishes under this new light. There is no reason why a person should prefer to gamble his or her fortune away, in all probability, simply because the expected monetary value of the gamble is infinite, rather than to stand pat.

Modern utility theory has another important consequence. It enables numerical utilities, corresponding to degrees of preference, to be assigned in games whose outcomes are inherently qualitative. The method of quantifying preferences among such outcomes on an interval scale of utilities can be sketched with the help of the game of Angelo and Isabella which was discussed in Chapter 1. For each player, we can begin by assigning arbitrary utilities, say 1 and 10, to the least and most preferred outcomes respectively. (In other interval scales, such as the Fahrenheit and centigrade or Celsius scales of temperature, two values corresponding, for example, to the freezing and boiling points of water are also assigned perfectly arbitrarily.) It is then possible to determine the utilities of the other outcomes by finding lotteries involving the two extreme outcomes which the player considers equally desirable to each of these others. Let us consider one of the possible outcomes from Isabella's point of view: she preserves her chastity and her brother is executed humanely. It may turn out that Isabella is indifferent between this outcome and the following lottery involving her least and most preferred outcomes: A die will be cast, and if a six comes up she will surrender her chastity to Angelo and her brother will be tortured to death, while if any other number comes up she will preserve her chastity and her brother will be reprieved.

Isabella is indifferent between a certain outcome and a 1/6:5/6 lottery involving her least and most preferred outcomes. If the outcome she considers worst is arbitrarily assigned a utility of 1 and her favourite outcome a utility of 10, then the utility of the third outcome under consideration is

$$(1/6)(1) + (5/6)(10) = 8.5.$$

This number is much closer to 10 than to 1; this reflects the fact that the outcome is a relatively attractive one—compared with the worst and the best that can happen—from Isabella's point of view. Every possible outcome of the game can be assigned a numerical utility in this way, and the utilities correspond to Isabella's degrees of preference among the outcomes. The same can, of course, be done from Angelo's point of view, and provided that the players are able to express consistent preferences between lotteries and outcomes, the payoffs of any game can be expressed in terms of numerical utilities. Utilities defined in this way are measured on interval scales. This means that the zero points and the units of measurement are arbitrary— mathematically, utilities are unique up to a positive linear transformation— so that they can all be multiplied by a positive constant, and a constant can be added to each of them, without affecting the information which they contain. This information concerns *relative* preferences or *ratios* of *differences* between preferences. It is important to bear in mind that a player's relative preferences among the outcomes, as they are reflected in the utilities, may be based on any factor(s) whatsoever that influence the player's degrees of satisfaction or dissatisfaction with the possible outcomes, including spiteful or altruistic attitudes towards the other player(s), religious beliefs, phobias, masochistic tendencies, and so forth.

Von Neumann and Morgenstern's theory of utility has not solved all problems concerning the assignment of numerical utilities to the outcomes of games, and some of the remaining difficulties will surface in later chapters. It does, on the other hand, provide a fairly convincing method of solving games of chance involving pure risk. It follows almost tautologically from the theory that in any such game a rational person will adopt the strategy that maximizes expected utility. When the outcomes are relatively small amounts of money, or can be readily interpreted as such, the principle of maximizing expected monetary value outlined in the previous section will generally yield solutions which closely approximate those generated by the modern utility principle, because utility in these cases will usually be a linear function of monetary value. But the utility principle is obviously much wider in scope than the older one.

Empirical research (see, e.g., Edwards and Tversky, 1967, and Becker and McClintock, 1967, for reviews of relevant experiments) has shown that utility functions can be constructed reasonably successfully for a wide variety of risky choices. Investigations of the strategies chosen by subjects in risky games of chance, however, have produced rather surprising results. In some very simple games of this type, subjects tend to choose strategies that are obviously non-optimal. Humphreys's (1939) light-guessing experiment is the classic in this field; a large number of later experiments have been modelled on it (e.g. Goodnow, 1955; Siegel and Goldstein, 1959; Siegel, Siegel, and Andrews, 1964; Ofshe and Ofshe, 1970). In these experiments the subjects are typically seated in front of two light bulbs and requested to predict which one will illuminate on each trial. The bulbs are illuminated in a random pattern according to probabilities fixed in advance; for example the left-hand bulb might be lit in 80 per cent and the right-hand bulb in 20 per cent of trials. Subjects typically begin by distributing their choices equally between the two options. After a number of trials they tend to increase the frequency of choosing the more frequently reinforced option. But after one or two hundred trials, by which time it is reasonable to assume that the probabilities are roughly known and a game of chance involving risk occurs on each trial, most subjects tend to settle into a non-optimal *matching strategy*. In other words, they choose strategies with probabilities approximately equal to those with which the bulbs illuminate, for example 0.8 left and 0.2 right in the case mentioned above. This is clearly non-optimal if the subject's utility for a correct guess is higher than that for an incorrect guess, which presumably it is, because the likelihood of a correct guess is maximized by choosing the more frequently illuminated bulb on every single trial. With the probabilities mentioned above, this optimal strategy yields 80 per cent correct guesses, while the matching strategy yields only a $(0.8)(0.8) + (0.2)(0.2)$ probability of being correct, that is, 68 per cent correct guesses. These and other similar experiments provide vivid illustrations of irrational behaviour in games of chance involving pure risk (see the reviews by Slovic, Fischhoff, and Lichtenstein, 1977; and Einhorn and Hogarth, 1981; and the discussion of recent experiments by Kahneman and Tversky, 1982).

2.5   Uncertainty

In everyday life, one often encounters one-person games against Nature that involve uncertainty rather than risk. In these cases the player not only does not know for certain what the outcomes of the available strategy choices will be, but cannot even assign meaningful probabilities to them. Uncertainty and risk both imply a degree of ignorance about the future, but the former is a profounder type of ignorance than the latter.

The following example is typical. A freelance journalist wishes to send an article through the post to a magazine in a foreign country. If the article is safely delivered, she will receive a fee of £200. It is possible, however, that the article will be lost in the post. The cost of ordinary postage is negligible, but for £5 she can register the parcel; if it is then lost, the £200 which she would have received from the magazine will be reimbursed by the Post Office.

If the journalist registers the parcel, then she loses £5 in postal registration charges, but receives £200 either from the Post Office (if the parcel is lost) or from the magazine (if it is safely delivered). If she neglects to register it, she receives nothing if it is lost, but £200 with no registration charge to pay if it is safely delivered. The journalist's net monetary gains in each of the four possible combinations of strategies of her own and Nature's can be most neatly summarized in what is called a payoff matrix (Matrix 2.1).

Matrix 2.1

		Nature	
		Lose	Deliver
Journalist	Register	195	195
	Don't register	0	200

It is possible, of course, that the figures in the payoff matrix do not accurately reflect the journalist's utilities. There may be other aspects of the outcomes apart from their monetary values which affect her preferences. Her reputation with the magazine for reliability, and her satisfaction at seeing the article in print may, for example, be worth more than the figures in the payoff matrix suggest. But it was shown in the previous section that considerations like these present no problems in principle: we may assume that the payoffs accurately reflect the journalist's utilities, or rather that they have been adjusted appropriately in order to do so.

What principles of rational choice can be offered in a case like this? If the journalist knows the probability of the parcel being lost, then the game is a straightforward risky decision, and the principle of rational choice outlined in the previous section is indicated. If, for example, she knows that the chances are one in ten that it will be lost, then the expected utility of registering it is

$$(1/10)(195) + (9/10)(195) = 195,$$

and the expected utility of not registering it is

$$(1/10)(0) + (9/10)(200) = 180;$$

so she would maximize by registering the parcel if she were to choose rationally in accordance with her preferences. But suppose that the journalist has no knowledge of Nature's probabilities; in other words she has not the faintest idea how likely it is that the parcel will be lost. The country for which it is destined, let us say, is going through a period of general industrial unrest and no one knows whether or not the postal services are affected, in fact no one can even hazard a guess. A number of principles of rational choice have been suggested for games involving uncertainty such as this.

(a) *Insufficient reason*

In practice, a decision involving uncertainty can always be transformed into a risky one by assigning probabilities to Nature's strategies arbitrarily. The principle of insufficient reason, advanced by the English clergyman Thomas Bayes in the middle of the eighteenth century, and later championed by Laplace and many others, is based on this idea. According to the principle, one is supposed to be justified in assigning equal probabilities to any events in the absence of any sufficient reason to believe that they are not equal.

The journalist in the example, using this principle, could assume that the probability of the parcel being lost is 1/2, and having made this assumption, the method of expected utility maximization outlined above could be applied. This method of solving—or perhaps one should say of side-stepping—the problem of uncertainty is, unfortunately, illusory. If the probabilities are completely unknown, then they are simply completely unknown; there is no logical justification for calling them equal. "Subjective probabilities", or degrees of belief, which are popular with many modern Bayesian statisticians, have not placed the principle of insufficient reason on a sound foundation. Some writers, for example Shepsle (1972), nonetheless believe that "*all* decision problems may be treated, subjectively at least, as risky choice" (p. 278, Shepsle's italics).

The principle of insufficient reason can be shown to lead to contradictions as follows. Suppose the journalist in the example develops a slightly different abstract model of the dilemma in which *three* possible states of Nature are incorporated: the parcel may be lost, delivered on time, or delayed. She would presumably be entitled to assign equal probabilities of 1/3 to each of these events. But we have already seen that she is entitled, according to the same principle, to assign the probability of 1/2 to one of them. By modelling the game in various other logically acceptable ways, any probability she might fancy could be assigned to the loss of the parcel. The principle of expected utility maximization, if used in conjunction with these fanciful probabilities, would yield different "solutions" depending on the model which was arbitrarily chosen. Treating uncertainty as though it were risk therefore seems to be a quite unacceptable approach to games of this sort.

(b) *Maximax*

This principle can hardly be considered rational, although it is often adopted by naïve decision makers such as young children and also by voters in elections (see Chapters 10 and 11). It counsels the player to choose the strategy that yields the best of the best possible outcomes. In the case of the journalist, the highest possible payoff if the parcel is registered is 195, and if it is not registered the highest possible payoff is 200. These figures are the maxima of the two rows of the payoff matrix. The maximum of the two maxima is 200, and it corresponds to the journalist's decision not to register the parcel. If she follows the maximax principle, therefore, she will neglect to register it.

The maximax principle amounts to angling for the most favourable outcome of the game while blissfully ignoring the less favourable outcomes which are possible. It is an ultra-optimistic approach to decisions under uncertainty since it is implicitly based on a hopeful anticipation of the highest conceivable payoff. Although it possesses a certain innocent charm, it is a transparently silly principle, and it does not normally characterize the behaviour of rational decision makers (but see Section 11.9).

(c) *Maximin*

Inspired by von Neumann and Morgenstern's (1944) classic text on game theory, the statistician Abraham Wald suggested this principle in 1945. The player is advised first to determine the worst payoff that can result from each possible decision, in other words to look for the minimum element in each row of the payoff matrix. The final step involves choosing the option or strategy that offers the best of these worst possible outcomes. Thus the player chooses the row that maximizes the minimum possible payoff; this row corresponds to the maximin strategy. In the case of the journalist, the maximin principle obviously leads to her choosing the strategy of registering the parcel, because the minimum payoff is then 195, compared with 0 if she does not register it.

In contrast to the maximax principle which is ultra-optimistic, the maximin principle is ultra-pessimistic. It is essentially a conservative method of maximizing security rather than of trying to get the most out of the game by taking chances. This can be seen most clearly in a payoff matrix such as Matrix 2.2. According to the maximin principle, the player should in this

Matrix 2.2

	Nature	
	0	1,000
Player		
	1	1

case choose the strategy corresponding to the second row, which guarantees a minimum payoff of one unit, rather than the first, which risks a zero payoff but offers the possibility of 1000 units. In fact, no matter how large the top right-hand matrix element might be, provided that the others are unchanged, a player guided by the maximin principle will never choose the first row.

This pessimistic policy of expecting the worst may be quite sensible when a great deal is at stake, and where the player's prime consideration is to avoid bad outcomes as far as possible at any cost. It is the principle adopted in modern statistical decision making, in which the statistician is more concerned with avoiding a Type I error (concluding that an effect is "real" when it is in fact due to chance) than a Type II error (concluding that it is due to chance when it is in fact "real"). But it does not seem intuitively to be a reasonable principle for *all* decisions under uncertainty, as we shall now see.

To understand the social implications of unbending pessimism, consider the commonsense maxim "Nothing venture, nothing gain". An interpretation of this maxim in game theory terms can be illustrated by considering what would happen if the maximin principle were applied in industry to decide whether or not to devote funds to research on product development or new manufacturing processes. (The following analysis is based on an example given by Moore, 1980, pp. 284–6.) The options are to support the research or to neglect it. Nature's moves determine whether the research succeeds or fails if it is supported. If the cost of a research programme is represented by c and the potential return if it is successful by r, then the payoff matrix is as shown in Matrix 2.3.

Matrix 2.3

		Nature	
		Research succeeds	Research fails
Player	Support research	$r-c$	$-c$
	Neglect research	0	0

The minimum payoff if the research is supported (row 1) is clearly $-c$, and the minimum if it is neglected (row 2) is zero. The maximum of these minima is zero no matter what the values of r and c might be. Thus, by adopting the ultra-pessimistic maximin principle, industry would *never* support research with uncertain outcomes, no matter how small the cost and how profitable the potential results! This is absurdly unadventurous, and it highlights the limitations of the maximin principle. It is a logically unassailable principle in so far as it achieves a certain clearly specified goal, that of ensuring the best of the worst possible outcomes. While this goal may be a reasonable one to pursue in some circumstances, it does not correspond to intuitive notions of rationality in others.

(d) *Minimax regret*

This principle was originally advanced by Savage (1951). It is regarded by many theorists as the soundest method of solving one-person games involving uncertainty since it is neither excessively optimistic nor excessively pessimistic, but it contains what many consider to be a subtle flaw. Savage's ingenious idea is to begin by transforming the payoff matrix of the game into a *regret matrix* (sometimes called a *loss matrix*). Each element in the regret matrix represents the positive difference between a specific payoff and the highest payoff possible under the corresponding state of Nature. These new numbers reflect the player's "opportunity loss" or degree of regret—once Nature's choice is revealed—for not having chosen the strategy that yields the highest payoff in those circumstances. Each payoff is converted into an element of the regret matrix by asking: Could the player have obtained a higher payoff by choosing differently if Nature's choice were known in advance? If not, then the corresponding regret is zero. If the answer is Yes, then the regret is equal to the amount by which the player could have improved the payoff by choosing the most profitable option given advance knowledge of Nature's choice.

To illustrate these ideas, let us reconsider the case of the journalist introduced earlier, and suppose that she chose to register the parcel. If it is delivered safely, her regret for having registered it amounts to 5 units, since she could have achieved a payoff 5 units higher by choosing not to register it under the circumstances. If the parcel is lost, on the other hand, she has no cause for regret since she could not have done better by not registering it, given this state of Nature. Arithmetically, each element of the regret matrix is found by subtracting the corresponding payoff from the highest in its column (Matrix 2.4).

Matrix 2.4

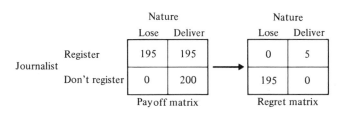

		Nature			Nature	
		Lose	Deliver		Lose	Deliver
Journalist	Register	195	195		0	5
	Don't register	0	200		195	0
		Payoff matrix			Regret matrix	

Notice that every element of the regret matrix is either positive or zero. This will always be the case, even if some or all of the original payoffs are negative: there is no such thing as negative regret! The minimax regret principle suggests the following method of solving one-person games involving uncertain outcomes: Choose the option that minimizes the maximum possible regret. If the journalist registers the parcel, then the maximum possible regret (in the top row of the regret matrix) is 5 units, and if she does not register it, then the maximum possible regret (in the bottom row) is 195 units. According to this principle, she will therefore register the parcel if she is rational.

It is worth examining the minimax regret solution to the game of research and development mentioned in connection with the maximin principle. The payoff matrix is reproduced below, together with its corresponding regret matrix. The regret matrix was calculated in the usual way, with the assumption that the cost of research (c) is less than the return if it succeeds (r) (Matrix 2.5). Now the player is counselled to examine the maximum of each

Matrix 2.5

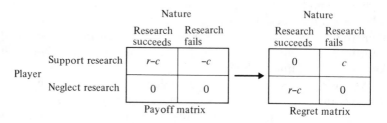

row of the regret matrix and to choose the row containing the least of these maxima. The maximum of the top row of the regret matrix is c, and that of the second row is $r - c$. The player should therefore support the research if c is less than $r - c$, which is equivalent to c being less than $r/2$, and neglect research otherwise. In other words, the player should support research only if the cost is less than half the potential return. This rule-of-thumb is quite commonly used in research management (Moore, 1980, p. 286), although it is normally based on intuition and experience rather than game theory analysis.

Unfortunately, the minimax regret principle suffers from what some theorists (e.g. Chernoff, 1954) consider to be a terminal malady: in some cases it violates a condition of rationality known as the *independence of irrelevant alternatives*. (This condition reappears in a modified form in Chapter 10 in connection with voting games.) The following example is based on one given by Luce and Raiffa (1957, p. 288). Suppose a customer in a restaurant consults a menu and narrows the choice down to either broiled salmon or sirloin steak. Eventually, the customer decides to order the salmon. The waiter then announces that, although it is not listed on the menu, roast chicken is also available. "In that case", says the customer, "I'll have the steak." This behaviour seems irrational, because the availability of chicken is irrelevant to the customer's preference between salmon and steak. But it can arise from following the minimax regret principle, as we can see by returning to the example of the journalist.

The minimax regret principle suggests that the journalist will register the parcel if she is rational. Suppose, however, that she contemplates a third strategy: she considers the possibility of registering the parcel and simultaneously betting a colleague £200 at even odds that it will be safely delivered. If the parcel is lost, she then loses the £5 postal registration charge and has to pay her colleague £200, but she has £200 reimbursed by the Post Office; thus she loses £5 overall. If the parcel is safely delivered, she loses the £5 registration charge, but is paid £200 by the magazine and wins another £200 from her colleague; her net gain is thus £395. The addition of this new hypothetical strategy enlarges the game without affecting the payoffs resulting from her original strategies, so that the new payoff matrix is the same as the old one with an additional row joined onto it. The new regret matrix can then be recalculated from it in the usual way. When the two possible payoffs associated with the "register and bet" strategy are joined onto the old payoff matrix, the new payoff matrix and its corresponding regret matrix are as shown in Matrix 2.6. The maximum in the first row of the regret matrix is

Matrix 2.6

	Nature Lose	Nature Deliver		Nature Lose	Nature Deliver
Register	195	195		0	200
Journalist Don't register	0	200		195	195
Register and bet	−5	395		200	0
	Payoff matrix			Regret matrix	

200, the maximum in the second row is 195, and the maximum in the third row is 200. The minimum of these maxima is 195, in other words the minimax regret strategy is "Don't register"! In the original game, the journalist's optimal strategy according to the minimax regret principle was to register the parcel, but as soon as an additional strategy was contemplated, the order of preference of the original strategies became reversed.

The above example is formally identical to the customer in the restaurant who orders salmon rather than steak but reverses this order when the alternative of chicken becomes available. If such a reversal seems irrational on the part of the restaurant customer, then it is equally irrational on the part of the journalist. The minimax regret principle is therefore regarded as suspect by many decision theorists.

One-person games in which the decision maker's choices are based on uncertainty are evidently more intractable than those that involve pure risk. When probabilities cannot be meaningfully assigned to the possible outcomes of the player's strategies, the principle of maximizing expected utility, which offers compelling solutions to risky decisions, cannot be used. Four principles which seek to define rational choices under uncertainty have been discussed, and they have all been found wanting. Others have been advanced from time to time, but none of them is entirely watertight, and it was proved many years ago (Milnor, 1954) that no principle could possibly satisfy all of the criteria which one would require for a completely convincing solution to the problem.

2.6   Summary

Section 2.1 drew attention to the existence of one-person games in which a solitary decision maker is pitted against Nature, and it was pointed out that analyses of these games are based on ideas that crop up in the solution of more complicated games involving two or more decision makers. Section 2.2 focused on decisions under certainty, and Section 2.3 on decisions involving pure risk. The superficially plausible principle of maximizing expected monetary value was seen to lead to difficulties, and in Section 2.4 it was shown how the worst of these difficulties can be overcome through von Neumann and Morgenstern's utility theory. A method was outlined of assigning numerical utilities on an interval scale of measurement to games whose outcomes are essentially qualitative in character. Section 2.5 was devoted to one-person decisions involving uncertainty rather than risk. Various suggested principles of solution—the principle of insufficient reason, maximax, maximin, and minimax regret—were examined, and all were found to violate intuitive notions of common sense in certain cases, although the last two were found to have some merits.

3

Pure Coordination Games and the Minimal Social Situation

3.1 Strategic Collaboration

THIS chapter is devoted to two classes of games involving pairs or groups of players whose interests are best served through strategic collaboration. The first is the class of pure coordination games characterized by perfect coincidence of the players' interests, and the second is a version of the minimal social situation which has received a great deal of attention from experimental investigators. The minimal social situation is a class of games involving incomplete information; in its strict form, the players are ignorant not only of the nature but even of the fact of their strategic interdependence.

In a pure coordination game, it is in every player's interest to try to anticipate the others' choices and obtain a mutually beneficial outcome, and they all know that the other players are similarly motivated. If the social situation makes unrestricted communication impossible, subtle and interesting psychological problems arise, and the favourableness of the outcome often depends upon the level of strategic intuition of the players. In the minimal social situation it is once again in the players' mutual interest to collaborate. The problems confronting the players are, however, quite different in this case, since they arise from a lack of information about the rules of the game. The central question in the minimal social situation is whether the players can learn, through playing the game a number of times, to collaborate without even being aware of each other's existence.

In the following section the general strategic properties of pure coordination games are examined. Theoretical and empirical work on the minimal social situation is discussed in Section 3.3, and a brief summary of the chapter is given in Section 3.4.

3.2 Pure Coordination Games

The defining property of a pure coordination game is complete agreement among the players about the order of preference of the possible outcomes: an outcome that is considered best by one player is considered best by the others, and the players' preference rankings of the other outcomes also coincide exactly. A consequence of this is that there is no conflict of interest between the players; their sole objective is to coordinate their strategies in such a way as to obtain an outcome that they all prefer. The game Head On, introduced in Chapter 1, is an example of a pure coordination game, since the outcomes are collisions or non-collisions and both players prefer the latter to the former. In Head On, each player chooses one of three pure strategies—swerve left, swerve right, or continue straight ahead—in an attempt to avoid colliding with the other in a corridor, and an outcome that is good (or bad) for one is similarly good (or bad) for the other.

Pure coordination games are clearly games of strategy according to the definition given in Chapter 1. In treating them as such, I am following Schelling's (1960) "reorientation of game theory", although many authorities do not regard them as genuinely strategic games. In their classic textbook, Luce and Raiffa (1957), for example, assert that any group of decision makers "which can be thought of as having a unitary interest motivating its decisions can be treated as an individual in the theory" (p. 13), and that if the players have the same preference pattern over the outcomes "then everything is trivial" (p. 59); "certainly in the extreme case where there is perfect agreement the analysis is trivial" (p. 88). There are, however, at least two good reasons for taking pure coordination games seriously. Although they may appear to be trivial from the point of view of formal game theory, informal analysis and empirical evidence discussed in this chapter shows that the problems confronting decision makers in these games are far from trivial. Secondly, it is desirable for aesthetic reasons to treat strictly cooperative payoff structures (that is, pure coordination games) as one extreme and strictly competitive structures (zero-sum games) as the opposite extreme, with mixed-motive games occupying the middle ground.

The theoretical symmetry between zero-sum and pure coordination games can be illustrated by a pair of simple examples which share certain features in common. Two-person Hide-and-Seek is a classic example of a zero-sum game: one player chooses from among a number of locations in which to hide and the other chooses which locations to search. Each outcome—discovery or non-discovery—represents a victory for one player and a defeat for the other, so the game is strictly competitive. If two people are accidentally separated in a crowded shopping centre, on the other hand, they must play a game which we may call Rendezvous. They must once again choose particular locations where they may or may not meet up again. But in this case each possible outcome is either good for both or bad for both, in other words they are playing a strictly cooperative pure coordination game.

Most parlour games are zero-sum, but Schelling (1960, p. 85) has pointed out that charades corresponds to a pure coordination game. Another example is the game "Mr and Mrs" played by married couples on British television. The husband and wife are each presented with the same set of questions about their marriage and neither is allowed to hear the other's answers; if their answers coincide, they win a substantial cash prize. A typical question which has been used is this: "If your kitchen needed redecorating, who would do the work?—The husband? the wife? the husband and wife together? a hired professional?". The options are presented to the players in different orders to prevent them from coordinating their answers artificially by agreeing in advance to choose the first (or the second, etc.).

A pure coordination game may involve just two players as in the above example or it may be a multi-person game, and the rules may specify perfect or imperfect information. The two-person, perfect information case is the simplest, and it presents no difficulties to the players. We recall from Chapter 1 that in a two-person game of perfect information the players move one at a time and each knows, when choosing a move, what move has been made by the other player. In Rendezvous, for example, if the players carry walkie-talkies, then one can easily inform the other of his or her initial move, thus making it a game of perfect information, and they can then meet up without difficulty; this is in fact how Rendezvous is solved by the police. Head On would have to be modelled in terms of perfect information if there were a convention or law governing behaviour in corridors, since each player would then know in advance which option the other would choose and successful coordination of strategies would be virtually assured; the "rule of the road" governing vehicular traffic exists for precisely this reason.

A multi-person pure coordination game of perfect information is not as straightforward as its two-person counterpart. The following example highlights the difficulties. During a revolution, soldiers are often faced with the problem of deciding whether or not to join the rebel army, and the majority decision may determine the success or failure of the revolution. Assuming that most of them do not care who wins but are anxious not to be on the losing side, their dilemma is aptly summarized in Benjamin Franklin's famous words: "We must all hang together or, most assuredly, we shall all hang separately." Even if the soldiers make their decisions one at a time in full knowledge of any previous decisions that have been made, in other words if the game is one of perfect information, the optimal strategy is not clear until it is evident which way the majority will decide. As more and more soldiers defect to the rebels or decide to remain loyal, a bandwagon effect is likely to develop. There is a psychologically critical point in the sequence of moves, long before a majority has come out in favour of either option, when the outcome seems "inevitable". From that point on the soldiers' optimal strategy is obvious since they know that the others will recognize this "inevitability" and choose accordingly. The expectations develop a motive power of their own which helps to determine the eventual outcome.

This type of self-fulfilling prophecy, which is called *tipping*, accounts for a variety of social phenomena which correspond to multi-person pure coordination games of perfect information. The spontaneous segregation of multi-racial cities into exclusively "white" and "black" areas when the critical ratio is reached has, for example, been analysed by Grodzins (1957). Schelling (1971a,b) has shown that extreme segregation is inevitable even if everyone prefers to live in a mixed neighbourhood provided that at least 40 per cent (say) of their immediate neighbours belong to their own group and they all change their places of residence in an attempt to achieve this goal. The same "invisible hand" governs the dissemination of vogues, fashions, and crazes.

Pure coordination games of imperfect information have quite different strategic properties. In these games, the players have to act in ignorance of one another's moves. They have to try to anticipate what the others will choose to do, and this involves anticipating the others' anticipations of their own choices. There is an endless regress of "I think that they think that I think . . . will choose *x*" inherent in the analysis. This may be illustrated in the television game Mr and Mrs which can be played at various levels of strategic depth. At the shallowest level the players choose their options according to non-strategic criteria. At a slightly deeper level the husband chooses the options that he thinks his wife will choose and vice versa. A deeper level than this involves the husband choosing the options that he thinks his wife will expect him to choose and vice versa. Even more profound levels of strategic thinking are, of course, possible in this apparently trivial game; in fact the scope for profundity is limitless.

The same strategic problem crops up in numerous everyday situations requiring strategic collaboration. The discovery of the endless regress created by imperfect information can be traced to a famous passage in John Maynard Keynes's (1936) *General Theory of Employment, Interest and Money*. Keynes discusses the strategy of buying and selling stocks and shares and compares it to a multi-person game of imperfect information with which most people are more familiar:

> Professional investment may be likened to those newspaper competitions in which the competitors have to pick out the prettiest faces from a hundred photographs, the prize being awarded to the competitor whose choice most nearly corresponds to the average preferences of the competitors as a whole; so that each competitor has to pick, not those faces which he himself finds prettiest, but those which he thinks likeliest to catch the fancy of the other competitors, all of whom are looking at the problem from the same point of view. It is not a case of choosing those

which, to the best of one's judgement, are really the prettiest, nor even those which average opinion genuinely thinks the prettiest. We have reached the third degree where we devote our intelligences to anticipating what average opinion expects the average opinion to be. And there are some, I believe, who practise the fourth, fifth and higher degrees (p. 156).

Some pure coordination games can be solved without difficulty by the players even if the rules prescribe imperfect information. This is always the case, for example, if (a) one particular combination of strategies produces an outcome that all players consider best, and (b) all players know the rules of the game and the preferences of the others (the condition of complete information). If these conditions are satisfied, the players can safely anticipate that the others will choose the strategies that produce the optimal outcome and that they will expect one another to do the same. In other cases, such as those that have been outlined in this section, no uniquely optimal combinations of strategies exist, and solutions must be sought through informal analyses based on factors outside of the abstract payoff structures of the games.

Schelling (1960) has shown that, in the absence of any uniquely optimal combination of strategies in a pure coordination game, informal analysis often reveals a *focal point* which possesses qualities of prominence or conspicuousness. If the players cannot communicate explicitly with one another, tacit coordination is often possible if each notices the focal point and assumes that the others will recognize it as the "obvious" solution. In Head On, for example, British players may recognize that for both to keep to the left is prominent because of the analogy with vehicular traffic, and if they are non-British keeping to the right may be prominent for the same reason. It is surprisingly difficult to think of any lifelike social situation corresponding to a pure coordination game that contains no clues pointing towards a focal point solution, even when the players have an infinity of options from which to choose. Tacit coordination which depends on shared cultural and psychological factors, is achieved via "telepathic" communication when explicit communication is infeasible. (The inverted commas serve to indicate that no supernatural phenomenon is intended by the phrase "telepathic" communication.)

The following pure coordination games, which were investigated empirically by Schelling (1960, chap. 3), illustrate the processes of tacit coordination and "telepathic" communication:

(1) Two people are invited to call "heads" or "tails" out of earshot of each other. If both call "heads" or both call "tails", then they both win; otherwise they both lose. Is there a sensible way to choose in this game? Is

either of the options prominent? Does an obvious focal point exist? Schelling found that 86 per cent of his subjects chose "heads" when confronted with this game. I have obtained even higher percentages in casual investigations in England, and also discovered that a much smaller percentage tends to call "heads" under non-strategic conditions. A person who normally calls "tails" to decide who will wash the dishes, for example, may nonetheless call "heads" in this game because of the realization that the other player will probably anticipate this choice and expect the first player to anticipate "heads" in turn, and so on. "Heads" is particularly prominent when the endless feedback of expectations is taken into consideration.

(2) Several people are assembled, and paper and pencils are distributed among them. Each person is invited to write down any amount of money. If everyone nominates the same amount, then they each receive this amount as a prize, otherwise none receives anything. In this game, each player has an infinite set of options from which to choose. Are any of these options prominent? Is this game soluble through informal reasoning? Schelling reported that 93 per cent of his subjects chose a "round sum" that was a power of 10, and no fewer than 30 per cent succeeded in converging on one million dollars. Choices other than powers of 10 included 64 dollars and 64,000 dollars which are culturally prominent "prizes for guessing" in the United States because of their use in radio and television quizzes. I have found British students to be comparatively unambitious in their choices, with a significant proportion modestly nominating the sum of £1.

(3) Two people who know nothing about each other are instructed to meet in New York at a particular time, but both know that no specific meeting place has been nominated. They would both like to keep the appointment, but they cannot get in touch with each other before the appointed time. What strategies would sensible players adopt in this awkward situation, which is merely a version of the familiar game, Rendezvous? Schelling reported that an absolute majority of his subjects, who were students in New Haven, Connecticut, managed to converge on the information booth at the Grand Central Station. The favourite choice of British subjects in an analogous game set in London is Piccadilly Circus. It is remarkable that people might succeed in coordinating in any large city, under the circumstances.

(4) The players in No. 3 are told the place and date but not the time of the rendezvous. Neither is prepared to hang about; they must both choose exactly the same time to within a minute. What time would people with sound strategic intuition choose to visit the meeting place? Despite the fact that each player has over a thousand options from which to choose

(since there are 1440 minutes in a day), Schelling found that virtually all his subjects chose 12 noon, and I have replicated this finding with British subjects. It is worth pointing out that 12 noon is prominent not merely on account of its astronomical significance but also for cultural reasons, since 12 midnight—an option which subjects never seem to choose—is no less significant according to purely astronomical criteria.

What is noteworthy about these results is the fact that players often succeed in coordinating their strategies against overwhelming odds in spite of the absence of formal solutions or prescriptions for choice based on logical reasoning. The ability to recognize the psychological prominence of certain combinations of strategies derives from a strategic intuition about how human beings are likely to interpret the problems, how they are likely to expect others to interpret them, and what expectations they are likely to hold about the expectations of others. It is difficult to imagine how a computer could ever be programmed to solved pure coordination games, because the method of analysis seems impossible to formalize. It takes a human strategist to anticipate the thinking of another human strategist in games like these.

3.3   The Minimal Social Situation

The minimal social situation, which was first investigated experimentally by Sidowski, Wyckoff, and Tabory (1956), is a class of games of imperfect information. In these games, the players are ignorant of their own payoff functions and those of the other player(s); each player knows merely what options may be chosen. In the extreme case, which I shall call the *strictly minimal social situation*, the players are not even aware of one another's existence: they know that they are making decisions under uncertainty, but they do not know that the uncertainty arises from their involvement in a game of strategy. In a situation such as this, the interesting possibility arises that collaborative behaviour may develop without the players realizing that there is anyone with whom to collaborate.

The following example illustrates the type of interdependent decision making that can be modelled by a minimal social situation. Two data analysts log in to a single computer from terminals in different buildings at the same time every day. Their activities are capable of disrupting each other's work, but they are oblivious of this fact and of each other's existence: the social situation is strictly minimal. Both wish to perform the same kind of computations and both know that there is a choice of two computer programs suitable for this purpose. Neither analyst knows, however, that the programs differ significantly in the demands they make on the computer's memory. A consequence of this is that if either analyst uses the limited-memory program, which they know simply as *L*, the work of the other is unaffected, but if either uses the redundant-memory program, *R*, then the work of the other analyst suffers severe delays. Delays at either terminal thus depend solely on the *R* program being used at the other, although the analysts are likely to assume that they are due to the unreliability of the program they themselves have chosen. The options available to the analysts are *L* and *R*, and each outcome involves a positive payoff (no delay) or a negative payoff (a delay) to each analyst. The payoff structure of the game is shown in Matrix 3.1.

Matrix 3.1
Mutual fate control

	L	R
L	+, +	−, +
R	+, −	−, −

It is unnecessary to attach any numerical significance to the matrix elements; we need to assume only that each analyst prefers no delay (+) to a delay (−). One analyst chooses row *L* or row *R*, and the other chooses column *L* or column *R*. The first symbol in each cell represents the payoff to the row chooser, and the second is the column chooser's payoff. If both choose the limited-memory program *L*, then the outcome is the upper-left cell; both analysts receive positive payoffs (there are no delays). If both choose *R*, then both suffer delays. If one chooses *L* and the other *R*, then only the former experiences a delay. The players' payoffs in this game are quite unaffected by their own strategy choices; their destinies are entirely in the hands of their partners, whose existence they do not even recognize. This type of payoff structure, referred to as *mutual fate control* by Thibaut and Kelley (1959), has most commonly been used in research on the minimal social situation. Although it is not a pure coordination game (some outcomes are favourable for one player and unfavourable for the other), it obviously calls for strategic collaboration: the solution is for both players to choose *L* every time the game is played.

Empirical research using this payoff structure under conditions of incomplete information was pioneered by Sidowski, Wyckoff, and Tabory (1956) and Sidowski (1957), who coined the term "minimal social situation". Subsequent experiments were reported by Kelley *et al.* (1962); Rabinowitz, Kelley, and Rosenblatt (1966), and others. The earliest experiments were devoted to the strictly minimal social situation, and in later experiments the information conditions have sometimes been partly relaxed.

In the original experiments of the Sidowski group, the appropriate conditions of strategic interdependence were engineered as follows. Pairs of subjects were seated in separate rooms, unaware of each other's existence, and electrodes were attached to their bodies. Each subject faced an apparatus on which was mounted a pair of buttons labelled L and R respectively and a digital display showing the cumulative number of points scored. The subjects were instructed to press one button at a time as often as they wished with the goal of obtaining rewards (points) and avoiding punishments (electric shocks). The rewards and punishments were arranged according to the mutual fate control payoff structure shown in Matrix 3.1. Whenever either of the subjects pressed the button labelled L, the other was rewarded with points; and when either pressed R, the other was punished with electric shock. (Less sadistic punishments, such as loss of points, have been used in more recent investigations.) From a game theory point of view, the situation is identical to that of the data analysts outlined above.

Sidowski, Wyckoff, and Tabory (1956) and Sidowski (1957) succeeded in demonstrating that, when the game is repeated many times, pairs of subjects generally learn to coordinate their choices although they are unaware of their strategic interdependence. They usually assume that their strategy choices are connected in some obscure way with their own payoffs, and the frequency of mutually rewarding outcomes tends gradually to increase over time. In the long run, both subjects normally settle down to choosing L on every occasion. The subjects behave as if they were learning to cooperate, although as far as they are concerned the situation is entirely non-social. How can these results be accounted for? Kelley *et al.* (1962) proposd a simple principle of rational choice for games of incomplete information which can, in principle, explain the phenomenon. It is based loosely on the well-known law of effect in psychology, and is called the *win–stay, lose–change* principle.

Thorndike's (1911) original statement of the law of effect was as follows:

> Of several responses made to the same situation, those which are accompanied or closely followed by satisfaction [are] more firmly connected with the situation . . .; those which are accompanied or closely followed by discomfort . . . have their connections with the situation weakened (p. 244).

In the minimal social situation, a person who adopts the win–stay, lose–change principle may be thought of as conforming to a version of Thorndike's law of effect. The principle does not generate any prediction about which options the players will choose on the first trial. If the game is repeated a number of times, however, the prediction is that each subject will repeat any strategy choice that is followed by a positive payoff and switch to the other after receiving a negative payoff. On the face of it, this seems to be an eminently sensible policy for a player to adopt in a game of incomplete information—such as the minimal social situation—which is played a number of times in succession. And this principle, in spite of its almost childlike simplicity, not only successfully accounts for the development of strategic collaboration in the minimal social situation, but also generates several non-obvious predictions which can be tested experimentally.

In Sidowski, Wyckoff, and Tabory (1956) and Sidowski's (1957) original experiments, the subjects were free to press their buttons whenever they wished. The following theoretical analyses of Kelley *et al.* (1962) are applicable to experiments in which the subjects are required to choose either simultaneously or alternately. Let us examine the case of simultaneous choices first.

Since the players' choices on the first trial are unpredictable, there are three initial outcomes to be considered. First, both players may initially choose L, with the result that both receive positive payoffs. If each adopts the win–stay, lose–change principle, they will then both repeat the same choice indefinitely:

$$LL \rightarrow LL \rightarrow LL \rightarrow LL. \ldots$$

Secondly, both may initially choose R. Since they both receive negative payoffs, both will switch to L on the second trial. The payoffs will then be positive, so both will repeat the same choice on all subsequent trials:

$$RR \rightarrow LL \rightarrow LL \rightarrow LL. \ldots$$

Finally, one player may initially choose L while the other chooses R. The first outcome (LR or RL) will then involve a positive payoff to one player and a negative payoff to the other. Following the win–stay, lose–change principle, the player who is rewarded (the one who chooses R) will repeat the same choice on the second trial, and the other will switch to R. The outcome on the second trial will therefore involve negative payoffs to both players, and we have reached the beginning of the series shown above. The complete analysis of this case is as follows:

$$LR \rightarrow RR \rightarrow LL \rightarrow LL \ldots$$
or $\qquad RL \rightarrow RR \rightarrow LL \rightarrow LL. \ldots$

The players will thus converge on the mutually rewarding combination of strategies in three trials and repeat these choices on all subsequent trials.

The theoretical analysis of simultaneous choices under the win–stay, lose–change principle has established, somewhat surprisingly, that successful coordination of strategies occurs within three trials at most, and as soon as the mutually rewarding strategies are once chosen simultaneously, the players continue to repeat them indefinitely. A further unexpected or counterintuitive conclusion of the theoretical analysis is this: except in the relatively unlikely event that the players hit upon the LL combination on the very first trial, they are bound to pass through a mutually punishing outcome (*RR*) before they can obtain a mutually rewarding one (*LL*). They have to hurt each other simultaneously before they can help each other simultaneously, unless they are lucky enough to help each other at the very outset. This suggests a possible strategic basis to the common belief in the criminal underworld that two people cannot develop a relationship of mutual trust until they have once quarrelled or fought with each other: an unpleasant conflict may "clear the air" for more harmonious interaction. The idea that conflict can help to promote intimacy has been advanced from a different theoretical perspective by Braiker and Kelley (1979).

Quite different conclusions flow from the analysis of alternating choices under the win–stay, lose–change principle. In the alternating procedure, the players move one at a time. One player chooses *L* or *R*, thereby delivering a reward or punishment to the other, and only then does the second player respond by choosing *L* or *R*, effectively rewarding or punishing the first. A second choice is then made by the first player, and this is followed by a second choice by the second player, and so on. The initial choices of both players must again be assumed to be arbitrary, but all succeeding choices can be determined according to the win–stay, lose–change principle.

If the first player initially chooses *L*, rewarding the second, and the second player also chooses *L*, rewarding the first, then the sequence will resemble the analogous process under the simultaneous procedure:

$$L \to L / \to L \to L / \to L \to L / \dots$$

If both initially choose *R*, however, the consequences are quite different. The first player punishes the second, and is then punished in return. The first player therefore switches to *L*, rewarding the second player, who therefore repeats his previous choice, *R*, punishing the first. The first player therefore switches back to *R*, punishing the second, and the second accordingly switches to *L*, rewarding the first. The first player then chooses *R* once again, punishing the second, and the second responds by switching to *R*. We have returned after four exchanges to the pair of choices with which the sequence began. This means that the whole cycle is bound to repeat itself indefinitely without ever generating a single consecutive pair of rewarding choices. The complete analysis is summarized below:

$$R \to R / \to L \to R / \to R \to L / \to R \to R / \dots$$

If the first player initially chooses *L* and the second *R* or vice versa, the reader can easily verify that the sequences will be as follows:

$$L \to R / \to R \to L / \to R \to R / \to L \to R / \dots$$
or $$R \to L / \to R \to R / \to L \to R / \to R \to L / \dots$$

Once again, the sequences of choices cycle indefinitely without ever producing a consecutive pair of rewarding choices. The cycles are in fact identical to the one initiated by a pair of R choices; they simply begin at different points in the same endless loop.

The theoretical analysis of alternating choices has produced results which contrast sharply with those pertaining to simultaneous choices. When the choices are made simultaneously, the win–stay, lose–change principle leads inexorably to mutually rewarding coordination of strategies within a very few trials. But when the players choose alternately, the same principle never leads to a coordinated solution, or even to transient mutual reward, unless the players stumble by chance on the rewarding strategies at the very beginning. In the light of these conclusions, the win–stay, lose–change principle loses some of its intuitive lustre as a general rule of rational conduct in games of incomplete information. It would be difficult to formulate a principle that yields worse results in the cases we have just examined; even an absurd win–change, lose–stay principle fares better in two of the four cases, as the reader can readily check.

Several unambiguous and testable though curiously non-obvious predictions flow from the theoretical analyses outlined above. Assuming that the players' initial choices are arbitrary but that they thereafter adhere to the win–stay, lose–change principle, the first prediction is that rewarding (L) choices and successful coordination will occur more frequently under conditions of simultaneous choice than in the alternating procedure, and the assumptions seem justified according to the law of effect. Kelley *et al.* (1962) confirmed this prediction in an experiment specifically designed to allow the relevant comparison to be made. Under the alternation procedure, the frequency of rewarding choices was no greater than chance expectation, and showed no tendency to increase when the game was repeated over 140 trials. This is precisely what the theory predicts. Under the simultaneous procedure, in contrast, rewarding choices exceeded chance expectation after relatively few trials and continued to increase in frequency, reaching 75 per cent after about 100 repetitions. Stabilized coordination of mutually rewarding strategies was achieved by many pairs under the simultaneous procedure but by only two pairs under the alternating procedure. These results have been replicated by Rabinowitz, Kelley, and Rosenblatt (1966), who also reported results in line with theory from minimal social situations involving different payoff structures.

Although these findings strongly confirm the prediction mentioned in the previous paragraph, they also reveal that most pairs of subjects did not adhere rigidly to the win–stay, lose–change principle. This is clear from the fact that most pairs did not succeed in achieving stabilized mutual reward within the first few trials in the simultaneous choice procedure; successful coordination typically took numerous trials to develop.

A second prediction from the theoretical analyses is that, under the simultaneous choice procedure, rewarding strategies will be chosen simultaneously by both members of a pair only after a mutually punishing outcome, unless the very first outcome happens to be mutually rewarding. Kelley *et al.* (1962) presented evidence which tends to confirm this prediction. In their simultaneous choice condition, every occurrence of a long run of mutually rewarding outcomes was immediately preceded by a mutually punishing outcome.

A third prediction centres upon the difference between strictly and non-strictly minimal social situations. In many everyday situations, the players are aware of the fact, though not the nature, of their mutual interdependence: they are informed of each other's existence and they realize that their choices influence each other's payoffs, but they do not know the payoff structure of the game. If the players adhere to the win–stay, lose–change principle under these circumstances, then their choices will be exactly the same as those in the strictly minimal social situation. But if the players know that they are involved in a game of strategy, even though they do not know the payoff structure of this game, it may be easier for them to coordinate their strategy choices. Thibaut and Kelley (1959) predicted that the frequency of rewarding choices and successful coordination will be greater under these circumstances. The players may be able to guess the payoff structure of the game by "trying out" their own available options and inferring from their partners' responses which is rewarding and which punishing.

Kelley *et al.* (1962) therefore compared the behaviour of informed pairs of subjects (in a quasi-minimal social situation) with that of uninformed pairs (in a strictly minimal social situation). The results strongly confirmed the predictions: rewarding choices and mutually rewarding outcomes were much more frequent in informed pairs. When choices were made simultaneously the relative frequency of rewarding choices in informed pairs rose to 96 per cent after about 150 trials. Even under the alternating procedure, the frequency of rewarding choices gradually increased over trials and greatly exceeded chance expectation. The results under the alternating procedure are particularly impressive in view of the fact that rigid application of the win–stay, lose–change principle generates no more rewarding choices than chance expectation and shows no tendency towards improvement over time, as shown above.

It is appropriate to draw the threads of this section together by returning to the example of the data analysts working on the same computer every day. Since it was assumed that they work simultaneously at separate terminals and are oblivious of each other's existence, the following conclusions are justified by theory and empirical evidence. In the first place, it is very likely that after a number of days they would both have learned to choose the computer program which does not cause delays at the other terminal. Secondly, they would probably not achieve this mutually beneficial pattern of behaviour until they had once disrupted each other's work *on the same day*, unless they chanced upon the solution—against which the odds are three to one—on the very first day.

With slight modifications to the example, further conclusions can be drawn. We may assume that the data analysts work on alternate days of the week, and that the use of the redundant-memory program causes delays—by altering a crucial data file, for example—to the analyst who uses either program on the following day. Under these conditions, the analysts would probably *never* learn to coordinate their choices in a mutually beneficial manner—to use only the limited-memory program—unless they both happened to choose this program at their first attempts, against which the odds are again three to one. Whether they use the computer simultaneously or on alternate days, however, their chances of learning to coordinate their choices in a satisfactory manner would be greatly increased if they became aware of each other's existence and of the fact that a delay for one depends in some obscure way on a choice made by the other.

The conclusions in the previous two paragraphs are far from obvious, and some are even counterintuitive. On the basis of common sense alone, or—to put it differently—without the benefit of theoretical analysis, strategy choices and outcomes in the minimal social situation are difficult or impossible to anticipate. But elementary strategic analysis enables clear-cut predictions to be made, and experimental results have tended generally to confirm these predictions.

3.4   Summary

This chapter began with an introduction to two classes of games in which the players' interests are served by strategic collaboration. Section 3.2 focused on pure coordination games characterized by complete coincidence of the players' interests. Although many of these games elude formal analysis, informal game theory usually uncovers prominent strategies and focal points which allow players with well-developed strategic intuition to coordinate tacitly through "telepathic" communication. Such empirical evidence as there is strongly confirms the prediction that successful coordination will often occur even when there are no "logical" solutions. Section 3.3 was devoted to theory and experimental evidence concerning the minimal social situation, which was interpreted as a class of games of incomplete formation. In the strictly minimal social situation the players are unaware, not only of the payoff structure of the game, but even of each other's existence. Theory and experimental evidence nonetheless reveal that strategic collaboration can develop when these games are played a number of times in succession. The win–stay, lose–change principle, derived from the law of effect, permits unambiguous though non-obvious predictions to be made about choices in the minimal social situation, and experimental evidence has generally confirmed these predictions.

Theory and Empirical Evidence

4

Two-Person, Zero-Sum Games

4.1 Strictly Competitive Games

A ZERO-SUM game, as the term suggests, is one in which the payoffs to the players in any outcome add up to zero; what one player gains, the other(s) must necessarily lose. A zero-sum game, in other words, is a closed system within which nothing of value to the players is created or destroyed: utilities merely "change hands" when the game is played. If there are just two players, this means that their interests are diametrically opposed, because an outcome that is favourable for one is bound to be correspondingly unfavourable for the other and vice versa. Since one player can gain only at the expense of the other, there are no prospects of mutually profitable collaboration, and models of two-person zero-sum conflicts are therefore described as *strictly competitive games*. They have proved especially amenable to formal analysis, and the most significant contributions to mathematical game theory pertain to them.

A wide variety of economic, political, military, and interpersonal conflicts correspond to strictly competitive games. Most two-person sporting contests and indoor games are strictly competitive in the zero-sum sense, and examples from less frivolous spheres of life are not difficult to find. Two television networks competing for audiences, two politicians or political parties competing for votes, two armies competing for territory, or two parents competing for the custody of their child after a divorce may have diametrically opposed interests which can reasonably be modelled by a two-person, zero-sum game. It is a common mistake, however, to regard all competitive interactions as zero-sum, since in reality the protagonists' interests are seldom strictly opposed. In most wars, for example, an outcome involving mutual annihilation with no gains on either side represents a loss for both contenders. Wars are seldom zero-sum, although isolated battles may be, as I shall presently show. Strictly competitive conflicts certainly occur in everyday life, but mixed-motive interactions are undoubtedly more common.

There are two major classes of strictly competitive games. Those in which both players have a finite number of pure strategies are called *finite*, two-person, zero-sum games, and methods are available for solving them. A solution consists of a specification of a rational way in which each of the players should choose among their available strategies, and a payoff, known as the *value* of the game, which results if both are rational according to the dictates of the theory. *Infinite*, two-person, zero-sum games, in which at least one of the players has an infinity of pure strategies from which to choose, sometimes possess formal solutions in this sense and sometimes do not. These games are, however, rather rare, and they are not dealt with in detail in this chapter. A clear introduction to the mathematical theory of infinite games has been given by Owen (1968, chap. 4).

The essential ideas of two-person, zero-sum game theory are outlined in the following sections. Two ways of representing the abstract structures of such games are explained with concrete examples in Section 4.2. The following two sections centre on the fundamental ideas behind the minimax solution. Sections 4.5 and 4.6 are concerned with special techniques for solving games. In Section 4.7 the problem is confronted of games in which information about the players' preferences is incomplete or non-quantitative, and Section 4.8 contains a brief summary of the chapter.

4.2   Extensive and Normal Forms

An incident from the Second World War, known by military historians as the Battle of Bismarck Sea, can be modelled by the simplest type of two-person, zero-sum game. The following account is based on a classic paper by Haywood (1954).

In February 1943, during the critical phase of the struggle in the southwestern Pacific, the Allies received intelligence reports indicating that the Japanese were planning a troop and supply convoy to reinforce their army in New Guinea. The convoy could sail either north of the island of New Britain where rain and poor visibility were almost certain, or south of the island, where the weather would probably be fair. By either route, the trip would take 3 days. General Kenney, who controlled the Allied forces in the area, was ordered by the supreme commander, General MacArthur, to attack the convoy with the objective of inflicting maximum destruction. General Kenney had to decide whether to concentrate the bulk of his reconnaissance aircraft on the northern or the southern route. Once the convoy was sighted, it would be bombed continuously until its arrival in New Guinea.

The players in this example were General Kenney and the Japanese commander. The options from which each had to choose were the northern and southern routes. The outcomes were the numbers of days of bombing resulting from each possible combination of choices. Kenney's staff estimated that if the reconnaissance aircraft were concentrated mainly on the northern route, then the convoy would probably be sighted after 1 day whether it sailed north or south, and would therefore be subjected to 2 days of bombing in either case. If the aircraft were concentrated mainly on the southern route, on the other hand, then either 1 or 3 days of bombing would result depending on whether the Japanese sailed north or south respectively. The number of days of bombing may be interpreted as Kenney's gains and the Japanese commander's losses. Since the Japanese payoffs are just the negatives of Kenney's, the corresponding game is obviously zero-sum. All of the essential information is summarized in Figure 4.1.

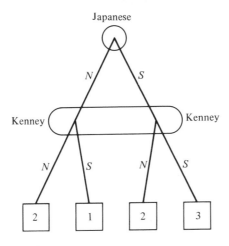

FIG. 4.1   Extensive form of the Battle of Bismarck Sea game.

In Figure 4.1 the *extensive form* of the game is represented in the most convenient way, by means of a *game tree*. The vertices of the game tree correspond to choice points and are labelled with the names of the players to whom they belong, while the branches represent the options between which the players choose. The initial (topmost) vertex is labelled "Japanese", indicating that the Japanese commander has the first move, and the succeeding vertices are labelled "Kenney". The terminal points of the game tree represent the outcomes which are reached after each player has moved in accordance with the rules of the game. If both choose south, for example, the outcome is 3 days of bombing as shown at the right-hand terminal point.

The vertices are enclosed in loops to indicate the *information sets* to which they belong. When making a move, a player cannot distinguish between choice points enclosed within a single information set. The initial vertex in Figure 4.1 is in an information set all on its own, but both of the vertices labelled "Kenney" are enclosed in a single information set. The purpose of this is to show that, at the time of choosing, Kenney does not know whether the left- or the right-hand vertex has been reached; in other words he does not know whether the Japanese commander has chosen to sail north or south. This means that the game is one of imperfect information—the players move simultaneously or in ignorance of any preceding moves—and it could have been portrayed equally well with the initial vertex labelled "Kenney" and the succeeding "Japanese" vertices enclosed in a single information set.

If the game were one of perfect information, on the other hand, then every vertex would be enclosed in a separate information set, like the initial vertex in Figure 4.1, and the sequence of moves represented in the game tree would assume significance. The extensive form of a perfect-information game such as chess, for example, would have to be represented in this way, although in practice it is impossible to do this for chess because the number of branches runs to billions after only a few moves. It is nonetheless possible to *imagine* a tree representing the extensive form of any finite game even if it cannot be drawn, and this is all that is required for some purposes. For a clear introduction to game trees, with several examples, see Singleton and Tyndall (1974, chaps 1 and 2).

A more compact method of representing a finite, strictly competitive game is by means of a rectangular array of numbers, called a *payoff matrix*, which displays the *normal form* of the game. Each row corresponds to one of Player I's pure strategies, and each column to one of Player II's pure strategies. Since a pure strategy is a complete plan of action specifying in advance how a player will move at any choice point which may arise in the course of play (see Chapter 1), the normal form allows a game involving a sequence of moves to be depicted statically, with the players simultaneously choosing a single row and column. The matrix element at the intersection of each row and column conventionally represents the payoff to Player I (the row chooser), and Player II's payoffs are simply negatives of those shown in the matrix. Any finite game in extensive form can be reduced to its equivalent normal form without loss of strategically relevant information. Matrix 4.1 shows the normal form of the Battle of Bismarck Sea game. The normal

Matrix 4.1
The Battle of Bismarck Sea

		Japanese	
		N	S
Kenney	N	2	2
	S	1	3

form of this game is extremely simple since each player chooses between only two options under conditions of imperfect information; the players' pure strategies are therefore simply their options. More complex examples are discussed later in this chapter.

4.3   Games With Saddle-Points

How might General Kenney, if he was rational, have analysed the game depicted in Matrix 4.1? At first glance, his optimal strategy seems to depend upon enemy intentions. If the Japanese decide to sail north of the island, then Kenney's best counter-strategy is to search north and obtain the outcome of 2 days of bombing. If they sail south, then Kenney does better by searching south and obtaining 3 days of bombing. But Kenney has no foreknowledge of the Japanese commander's intentions, so at this level of analysis his best counter-strategy remains indeterminate.

Rational generalship, however, is supposed to be based on enemy *capabilities* rather than enemy *intentions*. Kenney might therefore examine each of his available options from the viewpoint of the worst possible outcomes that could follow, given the options open to the Japanese. This pessimistic approach leads to the following conclusions: by searching north, Kenney is assured 2 days of bombing, whereas if he searches south the worst possible outcome is only 1 day of bombing. In other words, Kenney's minimum possible payoff if he chooses the northern strategy is 2 days, and the minimum possible if he chooses the southern strategy is 1 day of bombing. It follows that by choosing the northern strategy Kenney maximizes his minimum possible payoff. This choice is therefore called his *maximin* strategy; it has the property of ensuring the best of the worst possible outcomes. By choosing it, Kenney can guarantee that the payoff will not be less than two; this figure is his *maximin* value.

The Japanese commander can analyse the game in an analogous fashion. His objective is to minimize rather than to maximize the number of days of bombing. If he sails north, the worst that can happen—the maximum amount of bombing—is 2 days, and if he sails south, the worst possible outcome is 3 days of bombing. In order to ensure the best of the worst possible outcomes, he must therefore sail north. This choice corresponds to the Japanese *minimax* strategy since it minimizes the maximum possible payoff to the enemy. In this game, the minimum that the column player can guarantee (the minimax) is equal to the maximum that the row player can guarantee (the maximin): both are equal to 2 days of bombing.

According to the minimax principle of game theory, the optimal strategies available to the players in this type of zero-sum game are their maximin and minimax strategies. (For convenience, both are commonly referred to as minimax strategies.) In the Battle of Bismarck Sea game, players who are rational according to the minimax principle will choose their northern strategies, and the value of the game is 2 days of bombing since this is the payoff that results from rational play on both sides. These minimax strategies were in fact chosen by General Kenney and the Japanese commander in February 1943, and the Japanese suffered a disastrous defeat. The outcome cannot be attributed to any strategic error on the part of the Japanese commander; it was inherent in the unfair payoff structure of the game, whose value was positive and hence favourable to Kenney. In the event, the outcome would have been no better for the Japanese had they chosen differently, and Kenney, for his part, would have obtained a worse outcome (1 day of bombing instead of two) if he had chosen his non-minimax strategy.

The players' northern strategies are optimal because they intersect in an *equilibrium point* of the game. An equilibrium point is an outcome that gives neither player any cause to regret his choice of strategy when his opponent's choice is revealed. It represents a rational solution to any strictly competitive game for the following reason: it determines a minimax strategy for each player which yields the best possible payoff—the value of the game—against an opponent who also chooses optimally, and a payoff at least as good against one who chooses non-optimally. It should be borne in mind, however, that a minimax strategy does not necessarily exploit irrational play on the part of an opponent to maximum advantage. If there is good reason to suspect that the opponent will choose a non-minimax strategy, then a non-minimax choice may be the best counter-strategy. If, for example, Kenney suspected that the Japanese planned to sail south, then his best counter might be to search south although this choice would be irrational according to formal game theory. The minimax principle offers a persuasive definition of rational choice provided that there are no reasons for suspecting that one's adversary is irrational. Against an irrational adversary, however, it loses some of its force.

The equilibrium point of the Battle of Bismarck Sea game is easy to find because the payoff matrix contains a *saddle-point*. This technical term derives from the fact that a saddle is normally situated on a horse's back at the point which has minimum height on the animal's nose-to-tail axis and maximum height on its flank-to-flank axis. A saddle-point of a payoff matrix is an element that is a minimum in its row and a maximum in its column. The top left-hand element in Matrix 4.1 has this property, and it therefore corresponds to the equilibrium point of the game. Both of the numbers in row N are row minima because there is no number in this row less than 2. Only one of them, however, is also a column maximum, namely the left-hand one; therefore the NN cell is a unique saddle-point. If the payoff matrix of any strictly competitive game has a saddle-point, then its value is the value of the game and the players' minimax strategies are simply the corresponding row and column. If no saddle-point exists, then the minimum that Player I can guarantee by choosing a pure strategy—the maximin—is bound to be less than the maximum that II can guarantee, namely the minimax. If maximin and minimax are equal, and only then, a saddle-point exists in the payoff matrix. The fact that a saddle-point necessarily corresponds to the intersection of the players' minimax strategies is particularly useful for solving games that are more complicated than the Battle of Bismarck Sea game. This is illustrated in Matrix 4.2.

Since there is a saddle-point in Matrix 4.2, the minimax strategies and the value of the game can be found almost at a glance. By checking the minimum elements in each row to see whether any of them are also column maxima, the saddle-point is quickly located. It is situated at the intersection of row 2 and column 6. Player I's maximin is therefore equal to II's minimax, and each can guarantee, by choosing the appropriate pure strategy, that the payoff will be no worse from his own point of view than zero. The solution to this game is as follows: Player I should choose row 2, Player II should choose column 6, and the value of the game is zero (it is a fair game). If either player deviates from the prescribed minimax strategy while his opponent adheres to it, the "irrational" player will receive a payoff worse than the value of the

Matrix 4.2

II

	1	2	3	4	5	6	7	8
1	−1	−2	−3	−5	1	−2	2	1
2	1	3	2	2	1	0	2	1
3	−1	0	−3	−3	3	−4	2	−1
4	−2	−2	−6	−8	4	−2	4	0
5	0	3	−1	−1	4	−5	4	0
6	−2	2	−1	−2	3	−1	2	−1
7	−3	0	−4	−7	4	−3	4	0
8	−1	5	−1	0	4	−1	4	0

(Player I labels rows 1–8 on the left.)

game: Player I will be punished with one of the negative payoffs in column 6 if he alone deviates, and II's payoff following unilateral deviation will be the negative equivalent of one of the elements in row 2. In either case, of course, the minimax player will be rewarded with a corresponding gain beyond the value of the game. The minimax strategies are thus in equilibrium in the sense that the players would not regret choosing them if their opponents did likewise.

Unfortunately, not all payoff matrices have saddle-points. Games without saddle-points require more complicated methods of solution discussed later in this chapter. But a word needs first to be said about games with multiple saddle-points. A simple example of this kind is shown in Matrix 4.3.

Matrix 4.3

		II		
		1	2	3
I	1	4	2*	2*
	2	7	2*	2*
	3	3	0	1

All four of the matrix elements marked with asterisks in Matrix 4.3 correspond to saddle-points. Games of this kind present no new problems, however. It is easy to prove (see, for example, Luce and Raiffa, 1957, p. 66; or Singleton and Tyndall, 1974, pp. 47–51) that *all* of Player I's minimax strategies invariably intersect with *all* of II's, and the saddle-points are all of equal value. The solutions to the game shown in Matrix 4.3 are as follows: Player I should choose either row 1 or row 2 (both are optimal), II should choose either column 2 or column 3, and the value of the game is 2 units. All of the saddle-points are of course equilibrium points: a player who chooses a non-optimal strategy while his opponent chooses optimally will obtain a worse outcome than the value of the game, and his opponent will be rewarded by a corresponding amount.

4.4   Games Without Saddle-Points

The ideas discussed in the previous section cannot be applied without modification to all strictly competitive games. The following simple example from Sir Arthur Conan Doyle's story, *The Final Problem*, first subjected to game theory analysis by von Neumann and Morgenstern (1944, pp. 176–8), clarifies the difficulties:

> Sherlock Holmes desires to proceed from London to Dover and hence to the Continent in order to escape from Professor Moriarty who pursues him. Having boarded the train he observes, as the train pulls out, the appearance of Professor Moriarty on the platform. Sherlock Holmes takes it for granted—and in this he is assumed to be fully jus-

tified—that his adversary, who has seen him, might secure a special train and overtake him. Sherlock Holmes is faced with the alternative of going to Dover or of leaving the train at Canterbury, the only intermediate station. His adversary—whose intelligence is assumed to be fully adequate to visualize these possibilities—has the same choice. Both opponents must choose the place of their detrainment in ignorance of the other's corresponding decision. If, as a result of these measures, they should find themselves, *in fine*, on the same platform, Sherlock Holmes may with certainty expect to be killed by Moriarty. If Sherlock Holmes reaches Dover unharmed he can make good his escape (von Neumann and Morgenstern, 1944, p. 177).

Holmes's chances of escaping with his life are zero if he and Moriarty choose the same railway station. Let us assume further that his chances of survival are 100 per cent if he escapes via Dover, but only 50 per cent if he chooses Canterbury while Moriarty chooses Dover, since in the latter case the pursuit continues. It is reasonable to assume that Holmes's preferences are in correspondence with his chances of survival and that Moriarty's preferences are the negatives of these. If both players are aware of all these facts, then the payoff structure of the game is as shown in Matrix 4.4.

Matrix 4.4
Holmes v. Moriarty

Moriarty

		Canterbury	Dover
Holmes	Canterbury	0	50
	Dover	100	0

It is immediately obvious that this game has no saddle-point, since there is no matrix element that is the minimum in its row and the maximum in its column. Games without saddle-points are by no means uncommon. If a 2×2 (two-row, two-column) matrix is constructed at random, the probability that it will have a saddle-point is 0.67; in a 3×3 matrix the probability is 0.33; in a 4×4 matrix the probability is only 0.11; and in larger matrices the probability is extremely small. But payoff matrices are not random arrays of numbers, and it is reasonable to enquire whether there is any easily recognizable class of games whose matrices always have saddle-points. In 1912, Zermelo made the first major contribution to the theory of games of strategy by giving an answer to this question. Zermelo proved that the condition of perfect information is sufficient to ensure a saddle-point: every finite, strictly competitive game of perfect information has at least one saddle-point.

Zermelo's theorem, which is proved by reasoning backwards through the game tree, applies even to games that are far too complex to depict in extensive or normal form, let alone to solve. The payoff matrix of chess, for example, cannot be depicted because it contains billions upon billions of rows and columns, and the game tree cannot be drawn for a similar reason. But chess is a finite, strictly competitive game of perfect information, and it is therefore known to possess a saddle-point corresponding to optimal pure strategies for White and Black, although no-one knows what these optimal strategies are or even whether the value of the game is a win or a loss for White or a draw (see von Neumann and Morgenstern, 1944, pp. 124–5). An informal and simple proof of Zermelo's theorem can be found in Davis (1970, pp. 16–18).

Let us examine Holmes v. Moriarty, a game of imperfect information without a saddle-point, in an attempt to find a rational solution. If the players try to outguess each other, the analysis goes round in circles. The principle of choosing the pure strategy that ensures the best of the worst possible outcomes, which works with saddle-point games, also breaks down completely in this case. According to this principle, Holmes may choose either option, and the maximum payoff is zero. Moriarty, who wishes to minimize Holmes's payoff, should choose Dover, thereby holding the payoff down to a minimax of 50. But these machinations do not lead to a satisfactory solution for two reasons. The first is that the value of the game remains indeterminate since maximin is not equal to minimax; the most that can be said is that the value lies somewhere between 0 and 50. Secondly, if Moriarty's optimal strategy is indeed to choose Dover, then Holmes can anticipate this and counter by choosing Canterbury. At this point in the analysis, Moriarty's argument for choosing Dover collapses. The point is that Moriarty will be able to anticipate Holmes's choice of Canterbury, and will therefore be forced to reconsider the wisdom of choosing Dover. In the light of this, Holmes will also, of course, have to reconsider.

This conclusion is crippling to the whole argument, and it arises from the fact that the payoff matrix has no saddle-point. There is no combination of pure strategies that are in equilibrium, and consequently one of the players will always have cause to regret his choice when his opponent's is revealed. A security problem, unknown in saddle-point games, therefore exists: it is vitally important for each player to conceal his intentions from his opponent.

The surest way of concealing one's intentions is by leaving the decision to chance, and according to game theory that is precisely what a rational player in a non-saddle-point game ought to do! In order to ensure that his specific choice cannot be anticipated by his opponent, but that each option will have a predetermined *probability* of being chosen, a player can choose randomly by tossing a coin, casting dice, consulting a table of random numbers, or using some other suitable chance device. A player who behaves in this way is described in the terminology of game theory as adopting a *mixed strategy*.

It may seem paradoxical that a rational player should ever resort to a mixed strategy. A military commander, business executive, or political leader who was known to make important decisions by turning them into gambles would be regarded by many people to be in need of medical attention. (The diagnosis seems amply justified in some cases; see, for example, Luke Rhinehart's (1972) bizarre novel, *The Dice Man*.) But the rationality of randomized choice becomes obvious when extremely simple examples are examined. Consider the following game: Player I conceals a coin in one fist and II guesses "left" or "right". Player I wins one unit from II if the latter's guess is wrong, and pays II one unit if it is correct. Matrix 4.5 shows the payoff structure of this game.

Matrix 4.5

II

		L	R
I	L	−1	1
	R	1	−1

If either player chooses deliberately according to any principle whatever, he risks being "outguessed" by his opponent. This explains how adults, more often than not, manage to beat their children at this game. By choosing pure strategies, Player I can assure himself of a maximin of only −1 and Player II can guarantee to hold the payoff down to a minimax of no less than 1.

Suppose, however, that Player I uses a mixed strategy with equal probability assigned to left and right by tossing a coin, for example. The *specific* outcomes against each of II's pure strategies are then unknown, but their expected values can be calculated easily according to the principles outlined in connection with games of chance in Section 2.3. The expected payoff if II chooses left is $(1/2)(-1) + (1/2)(1) = 0$, and if II chooses right the expected payoff is $(1/2)(1) + (1/2)(-1) = 0$. The expected payoff also works out as zero against any mixed strategy that Player II might use. Player I's minimum expected payoff—the worst possible expected payoff—is therefore zero, which is considerably better than the maximin of −1 that he can guarantee through the use of a pure strategy. By using the mixed strategy, Player I cannot guarantee that he will not lose if the game is played only once, but he ensures that his chances are exactly even.

Similar considerations apply to Player II, who can hold the maximum expected payoff down to zero by using the same mixed strategy. The minimum and maximum expected payoffs, like the maximin and minimax payoffs in a saddle-point game, are therefore equal. The comparison can be carried further, since the mixed strategies are in equilibrium: neither player can profit by deviating from the prescribed mixed strategy if his opponent adheres to it. These strategies are therefore referred to as *minimax mixed strategies*, and they are considered to be optimal in the sense of being the best strategies available to rational players. Together with the value of the game, which is obviously zero since it is a fair game, they constitute the solution.

Von Neumann (1928) was the first to prove the fundamental minimax theorem, which establishes that *every* finite, strictly competitive game possesses an equilibrium point in mixed strategies. In the light of this theorem, saddle-point games are simply special cases involving degenerate mixed strategies in which probabilities of zero are assigned to every option but one. A simple proof of the minimax theorem is given in Appendix A. This theorem is the foundation stone of formal game theory, and any reader with at least school mathematics who wishes to acquire a true understanding of the theory is strongly urged to devote some time to studying it.

We are at last ready to provide a solution to the Holmes v. Moriarty game. Its payoff structure is reproduced in Matrix 4.6 with the optimal mixed strategies indicated in the right-hand and lower margins. Suppose that Holmes resolves to cast a die in order to choose the station at which he will leave the train. If 1, 2, 3, or 4 comes up (he decides in advance), he will alight at Canterbury, and if 5 or 6 comes up he will continue on to Dover. What are the expected payoffs against each of Moriarty's pure strategies? If Moriarty chooses Canterbury, Holmes's expected payoff is (2/3)(0) + (1/3)(100) = 33.33. If Moriarty chooses Dover, it is (2/3)(50) + (1/3)(0) = 33.33. Holmes's minimum expected payoff is therefore 33.33, which is a substantial improvement on the maximin of zero which results from his use of pure strategies only.

Matrix 4.6
Holmes v. Moriarty: Minimax solution

		Moriarty		
		Canterbury	Dover	
	Canterbury	0	50	2/3
Holmes				
	Dover	100	0	1/3
		1/3	2/3	

Suppose now that Moriarty uses a similar chance device to make his choice, but assigns a probability of 1/3 to Canterbury and 2/3 to Dover. The expected payoff if Holmes chooses Canterbury is (1/3)(0) + (2/3)(50) =

33.33, and the expected payoff against Holmes's choice of Dover is $(1/3)(100) + (2/3)(0) = 33.33$. The maximum expected payoff is therefore again 33.33, which is considerably better from Moriarty's point of view than the minimax of 50 which he can hold Holmes down to by using a pure strategy.

By using the indicated mixed strategies, Holmes can guarantee that the expected payoff will be no less than 33.33, and Moriarty can guarantee that it will be no more than 33.33. They are therefore minimax mixed strategies which are in equilibrium: neither player can obtain a better result by using any other (pure or mixed) strategy if his opponent plays optimally. The solution to the game is this: Holmes should choose randomly between Canterbury and Dover with corresponding probabilities of 2/3 and 1/3, Moriarty should randomize between the same options with probabilities of 1/3 and 2/3 respectively, and the value of the game is equivalent to a 33.33 per cent chance of Holmes's survival. In Sir Arthur Conan Doyle's story, Holmes alighted at Canterbury and watched triumphantly as Moriarty sped past on his way to Dover. This outcome was the one most likely to result from a combination of optimal mixed strategies, and from the novelist's point of view it had the virtue of allowing the harrowing pursuit to continue.

There is an important difference between 2×2 mixed-strategy games like the ones so far considered in this section and larger specimens. In the 2×2 case it is sufficient for *one* of the players to use a minimax mixed strategy to ensure that the expected payoff will be the value of the game; no more and no less. This means that a player is bound to obtain the same expected payoff by using any strategy, pure or mixed, against an opponent who plays optimally according to the theory. In larger mixed-strategy games, on the other hand, the value of the game is assured only if *both* players adopt optimal mixed strategies. In these cases a player who unilaterally deviates from minimax runs the risk of an expected payoff worse than the value of the game while his opponent benefits accordingly. The consequences of non-optimal play in non-saddle-point games may therefore be summarized as follows: There is never any advantage in using a non-optimal strategy against an opponent who is rational according to the minimax principle; but non-optimal play is positively disadvantageous only if the game is larger than 2×2.

The well-known Italian game, Morra, whose payoff matrix is 9×9, illustrates the full force of the minimax principle. The solution to this game, originally calculated by Williams (1966, pp. 163–5), is far from obvious (though it is easy to memorize), and a player who knows it is at a distinct advantage against an opponent who does not. The rules are as follows: each player extends one, two, or three fingers and at the same time calls out a number between one and three. The number called out is a guess regarding the number of fingers extended by the opponent, which is not known in advance since the players are required to move simultaneously. If only one player guesses correctly, then the loser pays the winner an amount corresponding to the *total* number of fingers displayed, otherwise the payoff is zero. If, for example, Player I extends one finger and calls out "one" (this pure strategy may be written 1–1), and Player II extends one finger and calls "three" (1–3), then the outcome is two units of payment to I, since only I's guess is correct and the total number of fingers extended is two. The complete payoff structure of this game is shown in Matrix 4.7.

Matrix 4.7 has no saddle-point, and the game therefore requires a solution in mixed strategies. The optimal mixed strategy is the same for both players, as is to be expected in a completely symmetrical game. The solution turns out to be this: Both players should ignore all options except 1–3, 2–2, and 3–1, randomizing among these three with probabilities of 5/12, 4/12, and 3/12 respectively, and the value of the game is zero (it is fair). A rational player might, for example, keep five red, four green, and three blue marbles in a convenient pocket. Before playing the game, he could then select one of the marbles at random and secretly examine its colour. Depending on whether it is red, green, or blue, he should then extend one finger and guess "three", extend two and guess "two", or extend three and guess "one". Against an opponent who uses the same mixed strategy, his chances of winning would then be even. Against one who uses any other (pure or mixed) strategy he may reasonably expect to win, and he would certainly come out on top in a sufficiently long series of repetitions.

Matrix 4.7
Morra

II

		1–1	1–2	1–3	2–1	2–2	2–3	3–1	3–2	3–3
	1–1	0	2	2	−3	0	0	−4	0	0
	1–2	−2	0	0	0	3	3	−4	0	0
	1–3	−2	0	0	−3	0	0	0	4	4
	2–1	3	0	3	0	−4	0	0	−5	0
I	2–2	0	−3	0	4	0	4	0	−5	0
	2–3	0	−3	0	0	−4	0	5	0	5
	3–1	4	4	0	0	0	−5	0	0	−6
	3–2	0	0	−4	5	5	0	0	0	−6
	3–3	0	0	−4	0	0	−5	6	6	0

In simple, naturally occurring strategic interactions, people sometimes use minimax mixed strategies intuitively, without conscious calculation. Davenport (1960), for example, analysed the strictly competitive game between Jamaican fishermen, who have to choose between fishing grounds, and their prey, which are controlled by currents; he showed that *both* players stick closely to the optimal mixed strategies prescribed by the minimax principle! The Jamaican fishing game has been reinterpreted by Kozelka (1969) and Walker (1977).

4.5 Dominance and Admissibility

One of a player's pure strategies is said to *dominate* another if it yields an outcome at least as good against any of the pure strategies that his opponent may choose, and a better outcome against at least one of them. In these circumstances he would clearly be acting irrationally if he chose the dominated strategy because it cannot produce a better outcome than the strategy which dominates it and may produce a worse one. A pure strategy that is dominated by another is *inadmissible* in the terminology of game theory, and a rational player will never choose it. A strategy that is not dominated by any other is admissible. The concepts of dominance and admissibility extend naturally to mixed strategies, but their practical usefulness is restricted mainly to pure strategies. The game depicted in Matrix 4.8 illustrates the basic ideas.

Matrix 4.8

	II	
	1	2
I 1	3	2
I 2	1	0

Row 1 clearly dominates row 2 because it leads to a better outcome for Player I whichever column II may choose. If II chooses column 1, then row 1 is better than row 2 because 3 is greater than 1, and if II chooses column 2, then it is better because 2 is greater than 0. Player I therefore has only one admissible strategy, namely row 1. Player II, whose payoffs are the negatives of those shown in the matrix, seeks to minimize his losses. Bearing this in mind, column 2 clearly dominates column 1 because 2 is less than 3 and 0 is less than 1. Hence column 2 is Player II's only admissible strategy.

The game is therefore solved: Player I should choose Strategy 1, Player II should choose Strategy 2, and the value of the game is 2 units. Not all strictly competitive games can be solved in this way, of course. The method cannot lead to a determinate solution unless there is a saddle-point in the payoff matrix, and even then it is not always applicable. Zermelo's theorem referred to in Section 4.3 above establishes, however, that if the game is one of perfect information—if the players move one at a time in full knowledge of all preceding moves—then not only will its corresponding matrix necessarily have a saddle-point, but the game will be soluble by the method of eliminating dominated rows and columns.

This method of solving games is not always quite as straightforward as the above example suggests; the elimination of dominated rows and columns sometimes has to proceed by successive stages. A slightly more complicated example is shown in Matrix 4.9. Row 2 is dominated by row 1, but no column appears to be dominated by any other. Since row 2 corresponds to a dominated strategy, however, a rational player will never choose it, so it is reasonable to delete it from the payoff matrix. *After deleting the dominated row* we are left with a 2 × 3 matrix in which column 3 is dominated by column 1. On the assumption that Player I will not choose row 2, therefore, Player II can safely disregard column 3. This assumption is reasonable since there are no circumstances in which I can do as well by choosing row 2 as by choosing another row.

Matrix 4.9

		II 1	2	3
I	1	3	6	4
	2	0	4	-2
	3	-2	-4	8

Following the same line of reasoning, the third column may also be deleted, leaving a 2 × 2 matrix in which only row 1 is undominated. The deletion of the row that is now dominated results in a 1 × 2 matrix whose first column dominates its second. Finally deleting the second column, we are left with a 1 × 1 matrix. Only the first row and the first column of the original matrix have survived the successive deletion of dominated strategies, and they intersect (as they are bound to) in a saddle-point. The solution is therefore for both players to choose their first pure strategies, and the value of the game is 3 units. Any game of perfect information can, in principle, be solved in this way, and so can many others. The solutions can usually, of course, be found more easily by simply locating the saddle-point, but the alternative method helps to clarify the assumptions underlying the solution. The dominance arguments can be applied to many games that are not strictly competitive, as will be shown in Chapters 6 and 8, and their practical usefulness in connection with strategic voting will be discussed in Chapter 11.

4.6   Methods for Finding Solutions

The minimax theorem asserts that every finite, strictly competitive game has an equilibrium point in pure or mixed strategies, but it offers no help in discovering this point in a specific case and thereby solving the game. There are, however, efficient methods for solving all such games, although some are quite difficult to apply.

The simplest case is a game with one or more saddle-points. When confronted with a game requiring solution, it is always wise to begin by examining its payoff matrix for saddle-points which, if they exist, allow a solution to be read off immediately. This method suffices for solving all games of perfect information and many others besides.

Games without saddle-points are almost as easy to solve as saddle-point games provided that each player has only two pure strategies. In the 2×2 case, a player's optimal mixed strategy yields the same expected payoff against each of the opponent's pure strategies, and this fact allows a straightforward method of solution. Player I's optimal mixed strategy may be written as a pair of probabilities x and $1 - x$, the first assigned to row 1 and the second to row 2. The expected payoffs yielded by this mixed strategy against column 1 and column 2 can then be expressed in terms of these probabilities. Since they are known to be equal, they can be equated, and the value of x can be found by solving the equation. Player II assigns probabilities of y and $1 - y$ to columns 1 and 2 respectively and obtains the same expected payoff from this mixed strategy against row 1 and row 2, hence the value of y can be found by solving an analogous equation.

In Matrix 4.4, for example, Player I assigns a probability of x to row 1 and $1 - x$ to row 2. His expected payoffs against column 1 and column 2 are equated as follows:

$$0x + 100(1 - x) = 50x + 0(1 - x),$$
$$x = 2/3.$$

Thus Holmes should assign a probability of 2/3 to row 1 (Canterbury) and $1 - 2/3 = 1/3$ to row 2 (Dover). From Moriarty's point of view

$$0y + 50(1 - y) = 100y + 0(1 - y),$$
$$y = 1/3.$$

Column 1 (Canterbury) should therefore occur with a probability of 1/3 in Moriarty's optimal mixed strategy, and Column 2 (Dover) with a probability of 2/3.

The value of the game is, by definition, the expected payoff when both players use optimal strategies. In the 2 × 2 case, however, this expected payoff is assured even if only one player uses an optimal mixture while the other chooses a pure strategy. The value of the game can therefore be found by substituting for x or y in any of the expressions on the left- or right-hand sides of the above equations. Suppose, for example, that Holmes uses his optimal mixed strategy while Moriarty chooses Canterbury. The expected payoff, and therefore the value of the game, is then $(0)(2/3) + (100)(1 - 2/3)$ or 33.33.

Some games that are larger than 2 × 2 can be solved by this method after the elimination of dominated rows and/or columns. An example is shown in Matrix 4.10. Column 2 is inadmissible; it is dominated by column 1 since Player II, who wants to minimize the payoff, does at least as well by choosing column 1 as by choosing column 2 against any of I's strategies. After deleting column 2, we are left with a 3 × 2 matrix in which row 1 turns out to be dominated by rows 2 and 3. Once row 1 has also been deleted, we are left with a 2 × 2 matrix. Further reduction through dominance is impossible and there is no saddle-point. The solution to the 2 × 2 sub-game can, however, be found by the methods described above, and the solution lifts back into the original 3 × 3 game. The deleted strategies—row 1 and column 2—should be assigned zero probabilities, and the optimal mixed strategies in the residual 2 × 2 game should be applied to the corresponding rows and columns of the original game. It turns out that Player I should mix Strategies 1, 2, and 3 with probabilities of 0, 1/4, and 3/4, and Player II should mix with corresponding probabilities of 5/8, 0, and 3/8. The value of the game is 13/4. If either player uses his forbidden row or column, or any mixed strategy that assigns a non-zero probability to it, his expected payoff will be worse than 13/4, provided of course that his adversary plays optimally.

Matrix 4.10

		II		
		1	2	3
	1	−3	5	−2
I	2	1	3	7
	3	4	4	2

A theorem proved by Shapley and Snow (1950) shows that any game in which *one* of the players has just two pure strategies can be solved by methods applicable to 2 × 2 games. The theorem asserts that one of the 2 × 2 sub-matrices embedded in the 2 × n or m × 2 payoff matrix has a solution that is also a solution to the larger game. To solve the original game, therefore, it is necessary only to solve each of its component 2 × 2 subgames and to check each of these solutions in the original game. If the player with only two pure strategies obtains an expected payoff from his prescribed mixed strategy against every one of his opponent's deleted pure strategies that is at least as favourable as the value of the 2 × 2 game, then a correct solution has been found. If the wrong 2 × 2 matrix has been used, the solution will not work in the larger game. An example is shown in Matrix 4.11.

Matrix 4.11

II

		1	2	3
I	1	−4	1	6
	2	9	0	−3

There is no saddle-point in this game, and it cannot be reduced by deleting dominated rows or columns. In view of the Shapley–Snow theorem, however, one of the 2 × 2 sub-matrices formed by arbitrarily deleting a single column must have a solution that lifts back into the original game. Let us try deleting column 3. We may solve the residual sub-game as follows:

$$-4x + 9(1 - x) = x + 0(1 - x),$$
$$x = 9/14.$$
$$-4y + (1 - y) = 9y + 0(1 - y),$$
$$y = 1/14.$$

Player I should therefore use row 1 with probability 9/14 and row 2 with probability 5/14, Player II should mix columns 1 and 2 with probabilities of 1/14 and 13/14 respectively, and the value of this 2 × 2 sub-game is −4(9/14) + 9(1 − 9/14) = 9/14.

If this is a solution to the original 2 × 3 game, then Player I's mixed strategy will yield an expected payoff at least as favourable as the value of the game against column 3; if it does not, we shall have to try out one of the other 2 × 2 sub-matrices. In fact, the expected payoff against column 3 is 6(9/14) − 3(5/14) = 39/14, which is clearly better for Player I than 9/14, so we need proceed no further; the mixed strategies and value already calculated are optimal in the original 2 × 3 game, and Player II should neglect his third pure strategy.

If a payoff matrix without a saddle-point has two rows and several columns, or if it has several rows and two columns, then the method described above will always yield a solution. The amount of labour involved may, however, be considerable if it contains a large number of embedded 2 × 2 sub-matrices. A matrix with two rows and eight columns, for example, contains twenty-eight 2 × 2 sub-matrices, all of which may have to be solved before a solution to the 2 × 8 game is found. In general, if the larger dimension of the matrix is n, the number of 2 × 2 sub-matrices embedded in it is $n(n - 1)/2$. Graphical methods which enable the critical 2 × 2 sub-matrices to be located quickly are, however, available (see, for example, Williams, 1966, pp. 71–81).

The solution of 3 × 3 and larger games, when none of the methods so far described applies, is extremely cumbersome. The most efficient procedure, which is applicable to all finite, strictly competitive games without exception, is based on the fact that the solution can always be reduced to that of a standard linear programming problem. The details of the *simplex algorithm*, as it is called, are beyond the scope of this chapter; the fundamental ideas are clearly explained in Singleton and Tyndall (1974, chaps 8, 9, 10, and 11), and the actual step-by-step procedure is painstakingly described in Williams (1966, chap. 6). I shall conclude this section by outlining a much simpler procedure, based on fictitious play, which enables solutions to be approximated with any desired degree of accuracy.

Imagine two players, both ignorant of game theory, playing the same game over and over again. Both are statistically inclined and keep records of the pure strategies chosen by their opponents. Being quite unable to analyse the matrix, their initial choices are arbitrary. On every subsequent occasion, however, each chooses the pure strategy that yields the best expected payoff against the opponent's mixed strategy, estimated on the basis of the relative frequency with which the opponent has thus far chosen each pure strategy. The consequences can, of course, be worked out on paper, and Julia Robinson (1951) proved that the players' choices are bound to converge towards optimal mixed strategies. Consider the example shown in Matrix 4.12.

Matrix 4.12

		II		
		1	2	3
	1	5	8	1
I	2	2	5	10
	3	9	0	5

Suppose the players arbitrarily choose row 1 and column 1 to start. Player I's second choice will be row 3, since this is best against column 1, and II's second choice will be column 3, which minimizes against row 1. Then I will again choose row 3 because it maximizes the expected payoff against II's 1/2, 1/2 mixture of columns 1 and 3, and II will again choose column 3 because it minimizes against I's 1/2, 1/2 mixture of rows 1 and 3. The process can be continued indefinitely with arbitrary choices being made whenever more than one pure strategy is best against the opponent's mixture. The following recipe simplifies the procedure:

(a) Copy row 1 and column 1 onto a sheet of paper. Column 1 may be written horizontally for convenience.
(b) Mark the minimum element in the row and the maximum element in the column with asterisks. Choose any minimum or maximum if there is more than one.
(c) Note the position of the asterisked column element and add the corresponding row to the first row. For example, if the second column element is asterisked, add row 2 to row 1, element by element. Note the position of the asterisked row element and add the corresponding column to the first column.
(d) Asterisk the minimum element in the resulting row total and the maximum element in the resulting column total. If the values of these two elements are sufficiently close to each other, then stop.
(e) Note the position of the column total element asterisked in (d), and add the corresponding row to the row total. Note the position of the row total element asterisked in (d) and add the corresponding column to the column total.
(f) Go to (d).

The recipe may be applied to Matrix 4.12 as shown in Table 4.1. The game has been played fictitiously 10 times. Each asterisk among the row totals indicates that Player II was about to choose the corresponding column on the following trial, and the asterisks among the column totals show which rows Player I was about to choose on the following trials.

Table 4.1

Row totals			Column totals		
col. 1	col. 2	col. 3	row 1	row 2	row 3
5	8	1*	5	2	9*
14	8	6*	6	12	14*
23	8*	11	7	22*	19
25	13*	21	15	27*	19
27	18*	31	23	32*	19
29	23*	41	31	37*	19
31	28*	51	39	42*	19
33*	33	61	47*	47	19
38	41*	62	52*	49	28
43*	49	63	60*	54	28

Counting the asterisks among the column totals we find that Player I chose rows 1, 2, and 3 with relative frequencies of 3/10, 5/10, and 2/10 or 0.3, 0.5, and 0.2 respectively. The asterisks among the row totals show that Player II chose columns 1, 2, and 3 with relative frequencies of 0.2, 0.6, and 0.2 respectively.

The bottom sets of figures show each player's total payoff after 10 trials against each pure strategy that his opponent might choose. Player I's worst possible total payoff, given the mixed strategy he in effect used, is 43 units, shown asterisked. This would have been Player I's total if II had chosen column 1 on every trial. The mixture used by Player I therefore yields a minimum possible payoff of 4.3 units per trial. The worst that could have happened to Player II is a total payoff of 60, which would have been the result if Player I had chosen row 1 on every trial as shown by the position of the last asterisk. Player II's mixture thus holds the maximum possible payoff down to 6.0 per trial.

We know from the minimax theorem, however, that the minimum which Player I can ensure by using an optimal mixed strategy is the same as the maximum which II can ensure by doing likewise, since both are equal to the value of the game. The above calculations show that the value of the game lies somewhere between 4.3 and 6.0. The difference between these two estimates is equivalent to 17 per cent of the range of payoffs in the game. It reflects the degree to which the strategy mixtures deviated from optimality, and a higher degree of accuracy can be obtained by repeating the fictitious play over more trials.

When the calculations are extended over 20 repetitions, the range of error decreases to the difference between a minimum of 4.5 and a maximum of 5.3, or 8 per cent of the payoff range. The relative frequencies with which Player I chooses Strategies 1, 2, and 3 become 0.5, 0.3, and 0.2 respectively, and those of Player II become 0.4, 0.3, and 0.3. These figures approximate the correct solution, which prescribes probabilities of 0.42, 0.33, and 0.25 for both players. The true value of the game is exactly 5 units.

The method of fictitious play can be used for finding an approximate solution to any finite, strictly competitive game. The approximation can, of course, be made as close as desired. It is the approved method of handling very large games.

4.7   Ordinal Payoffs and Incomplete Information

Throughout this chapter, the assumption has been made that the figures in the payoff matrices represent fully numerical utilities as defined in Chapter 1, in other words that the players' preferences are measured on interval scales. The condition of *complete information* has also been assumed throughout. This means that the players are assumed to have full and accurate knowledge of their payoff matrices. In everyday conflicts, however, the protagonists often lack such accurate and complete information about their opponents'—or even their own—degrees of preference. In this section I shall therefore investigate some of the consequences of relaxing the assumptions of interval-scale payoffs and complete information.

It is useful to establish first of all in what ways a payoff matrix can be modified without altering the strategic structure of the underlying game. As pointed out in Chapter 1, utilities are unique up to a positive linear transformation. This means that the strategic properties of a matrix remain the same if a constant is aded to each element or if each is multiplied by a positive constant. Consider Matrices 4.13.

The optimal strategies prescribed by the minimax principle are identical in all four matrices. In each case Player I should mix rows 1 and 2 with equal probabilities while Player II should mix columns 1 and 2 with probabilities of 1/4 and 3/4 respectively. The reason for this coincidence is that Matrices (b), (c), and (d) are positive linear transformations of (a). Matrix (b) is formed from (a) by multiplying each element by 12; (c) is formed by adding 8 to each element; and (d) is formed by multiplying each element by 1/7 and adding

Matrices 4.13

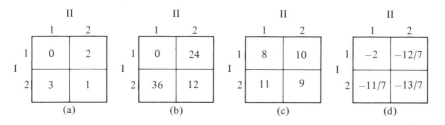

-2 to it. The values of the four games reflect these transformations: the value of (a) is 3/2, while those of (b), (c), and (d) are $(3/2)(12) = 18, 3/2 + 8 = 19/2$, and $(3/2)(1/7) - 2 = -25/14$ respectively.

Adding a constant to the payoffs of a game has the effect of altering the zero point of the utility scale on which they are measured, and multiplying by a positive constant merely changes the units of measurement. It is often convenient to transform a matrix in order to eliminate negative elements and fractions prior to solving it by hand. Whenever this is done, however, it is necessary to perform the opposite transformation on the resulting value of the game in order to bring it back in line with the original scale of measurement.

Some games can be solved on the basis of only fragmentary information about their corresponding payoff matrices. This is the case, for example, if the payoffs associated with just one row and one column of a matrix are known and they intersect in a saddle-point. An example of this is shown in Matrix 4.14. There is a saddle-point in this matrix corresponding to row 1 and column 1, and its existence is quite unaffected by the unknown payoffs in the other rows and columns. The players' first pure strategies are therefore optimal, and the value of the game is 1 unit. (Other solutions might of course emerge if more information about the payoff structure came to light, but in that case the values of the newly discovered saddle-points would be 1

Matrix 4.14

		II			
		1	2	3	4
	1	1	2	7	1
I	2	0	?	?	?
	3	−1	?	?	?
	4	0	?	?	?

unit as well.) This example shows how games with saddle-points can sometimes be solved by the players themselves or by game theorists who are ignorant of many of even most of the possible outcomes.

Many critics of game theory, and some game theorists, have drawn attention to the basic problem of measuring the players' utilities on an interval scale. Shubik (1964), for example, has had this to say:

> The numbers in the payoff matrix have to be obtained in some manner or the other. It is difficult enough to be able to state with certainty that an individual prefers to see Jones as a senator rather than Smith; it is more difficult (and some may say impossible) to state by how much he prefers Jones to Smith (p. 19).

The truth of the matter is that strictly competitive games can often be convincingly solved without any appeal to quantitative measurement. Any game in which the players move one at a time in full knowledge of all preceding moves—a game of perfect information—or any other saddle-point game for that matter, can be solved on the basis of purely ordinal preference rankings. In order to solve such a game it is not necessary to know the players' degrees of preference; one needs merely to know which outcomes the players consider best, second best, and so on. Brams (1980) has analysed many two-person conflicts from the Old Testament on the basis of ordinal payoffs. The following example, loosely based on one of Brams's models (pp. 132–9) shows how this is done.

Saul, the king of Israel, appointed David, "a mighty valiant man, and a man of war, and prudent in matters, and a comely person" (I Sam.xvi:18) in command of his soldiers. David set out with the army, slaughtered great numbers of Philistines, and was greeted on his homecoming by women dancing and singing "Saul hath slain his thousands, and David his ten thousands" (I Sam.xviii:7). Saul became jealous and afraid of David's growing popularity and made a number of attempts on his life. David was eventually persuaded to flee into exile.

Saul may reasonably be assumed to have had two options: to pursue David (P) or not to pursue him (DP). David, for his part, had two options: to flee Saul (F) or not to flee (DF). While it is impossible to assign accurate numerical utilities to the various possible outcomes of this game, certain reasonable assumptions about the payoffs can be made. To begin with, the conflict is evidently strictly competitive: an outcome that is good from Saul's point of view is correspondingly bad from David's, and vice versa. The game is therefore presumably zero-sum. Secondly, the outcomes can be arranged in order of preference. The best possible outcomes for Saul are those in which he pursues David and David does not flee, because in these cases David is killed and the threat is removed. Saul's worst possible outcomes are those in which he does not pursue David and David does not flee, since David's popularity is then likely to eclipse Saul's. Other possible outcomes are intermediate from Saul's point of view, as David remains in exile and the problem is unresolved. David's order of preference among the various possible outcomes is simply the reverse of Saul's.

Since David moves after Saul, and in full knowledge of Saul's move, the game is one of perfect information and must therefore have a saddle-point. This means that it will have a solution in pure strategies. In the normal form of the game, Saul's pure strategies are simply to pursue (P) or not to pursue (DP), but David, moving second, has four pure strategies:

F/F: flee whether pursued or not;
F/DF: flee if pursued, don't flee if not pursued;
DF/F: don't flee if pursued, flee if not pursued;
DF/DF: don't flee whether pursued or not.

The normal form of this game is displayed in Matrix 4.15, with the payoffs to Saul labelled g (good), f (fair), and b (bad).

Matrix 4.15
Saul and David

David

		F/F	F/DF	DF/F	DF/DF
Saul	P	f	f	g	g
	DP	f	b	f	b

A saddle-point is situated at the intersection of the first row and second column of the matrix. It can be identified as a saddle-point because "fair" is a minimum payoff in row 1 (it is obviously worse than "good") and a maximum in column 2 (it is better than "bad"). In fact the first row dominates the second since it leads to an equal or preferable payoff to Saul no matter which column is chosen by David, and the second column is dominant since it leads to outcomes equal or worse for Saul than any other column irrespective of Saul's choice. There is thus only one admissible strategy available to each player: Saul should pursue David, and David should flee only if he is pursued. The value of the game is "fair". The biblical account of what actually happened suggests that these optimal strategies were in fact chosen.

Games without saddle-points cannot be solved without interval-scale measurement of the players' preferences. Even in these games, however, a certain amount of progress can usually be made towards a solution on the basis of incomplete information. Let us return to the example of Holmes v. Moriarty discussed in Section 4.4 and assume that information about the players' preferences is incomplete.

Holmes and Moriarty can each choose to alight at either Canterbury or Dover. The former wishes to maximize his chances of survival and the latter to minimize them, but in this version of the game neither player can quantify these chances. Ordinal payoffs can readily be assigned, however. If both players choose the same station, then the outcome is "bad" for Holmes; if Holmes chooses Canterbury and Moriarty Dover, the outcome is "fair" for Holmes; and if Holmes chooses Dover and Moriarty Canterbury, the outcome is "good" for Holmes. Moriarty, of course, ranks the outcomes in the opposite order. The payoff structure is shown in Matrix 4.16.

Matrix 4.16
Holmes v. Moriarty: Ordinal payoffs

		Moriarty	
		Canterbury	Dover
Holmes	Canterbury	b	f
	Dover	g	b

There is no saddle-point in this ordinal game; the best payoff that Holmes can guarantee by using a pure strategy is "bad", and the worst that Moriarty can guarantee to hold it down to by doing likewise is "fair". The value of the game therefore lies somewhere between the "bad" maximin and the "fair" minimax; both players should therefore use mixed strategies.

Holmes's minimax mixed strategy can be worked out in the usual way by assigning a probability of x to Canterbury:

$$bx + g(1 - x) = f(x) + b(1 - x),$$

which yields $x = (b - g)/(2b - g - f)$. Holmes should assign this probability to Canterbury and $1 - x = 1 - (b - g)/(2b - g - f)$ to Dover. Simple algebraic manipulation reveals that the ratio of Canterbury to Dover in Holmes's optimal mixed strategy is simply $(g - b):(f - b)$. He should therefore assign a larger probability to Canterbury than to Dover because the difference between "good" and "bad" is larger than that between "fair" and "bad". Similar calculations show that Moriarty's optimal mixed strategy requires that he assign probabilities to Canterbury and Dover in the ratio $(f - b):(g - b)$, therefore he should assign a larger probability to Dover. Holmes's chances of survival if both players are rational according to the dictates of game theory are somewhere between "bad" and "fair", because, as indicated earlier, the value of the game lies in that range. Without exactly solving the game, we have succeeded in revealing a certain amount about its optimal mixed strategies and value.

4.8 Summary

The chapter began with a discussion of zero-sum games and an explanation of why two-person, zero-sum games, unlike any others, are strictly competitive. Formal game theory defines solutions—in terms of optimal ways of playing and a value which then results—to all conflicts of this kind, apart from some in which one or both of the players choose from an infinity of options. In Section 4.2 the main methods of representing the abstract structure of strictly competitive games—game trees and payoff matrices—were illustrated with examples. Sections 4.3 and 4.4 were devoted to the fundamental ideas behind the minimax solution, and in Section 4.5 the concepts of dominance and admissibility, which simplify the solutions of many games and clarify the logic of these solutions, were explained and illustrated. Section 4.6 centred on methods, based on saddle-points, dominance, simple probability calculations, and fictitious play, for solving finite, strictly competitive games of all varieties. In Section 4.7, cases in which information about the players' preferences is incomplete or non-quantitative were discussed, and it was shown how they can sometimes be solved or partly solved.

5

Experiments With Strictly Competitive Games

5.1 Ideas Behind Experimental Games

THE SUBTLE connection between game theory and experimental games is often misunderstood. Unlike most conventional theories in other fields of investigation, game theory is quite incapable of being empirically tested and falsified; in this respect it resembles geometry or probability theory rather than, say, psychophysics or relativity theory. The function of game theory is to provide an abstract framework for modelling situations involving interdependent choice. In some cases, including all finite, strictly competitive games, it helps us to discover how rational decision makers *ought* to behave in order to attain certain clearly specified goals; but about how people actually *do* behave it says nothing. The ways in which human beings typically respond to problems of interdependent choice is an essentially empirical question, and it is investigated by means of experimental games in which decision making is observed under controlled conditions.

In experimental games, the subjects are presented with decision tasks with game-like structural properties, and their choices are carefully recorded and analysed. A variety of decision tasks have been used in the long tradition of research in this area, but the most popular have undoubtedly been matrix games. In a conventional experimental matrix game, pairs of subjects are presented with a payoff matrix. One member of each pair is assigned the role of row-chooser and the other the role of column-chooser, and they are usually required to make a series of simultaneous choices. After each trial the subjects are awarded points, valueless tokens, or small amounts of money according to the payoff structure of the matrix.

Various modifications of this basic paradigm are possible. In some experiments the subjects are not fully informed about the payoff structure; in others they are pitted against opponents who respond with programmed sequences of choices; and sometimes the subjects are required to choose one at a time rather than simultaneously. The payoff matrix can even be dispensed with entirely: in negotiation games the subjects simply engage in verbal bargaining in an attempt to agree upon, for example, a price for an imaginary commodity which one must sell and the other must buy. Multi-person experimental games, discussed in Chapter 9, require slightly different techniques of investigation.

There are several obvious reasons for the enormous popularity of experimental games in the investigation of interdependent choice. The first and most important is the apparent ease with which the nature of the subjects' interdependence can be precisely specified with the help of concepts derived from game theory. The strategic structure of any two-person interaction that is likely to arise in everyday life can apparently be reproduced without difficulty in an experimental game via an appropriately chosen payoff matrix. Secondly, experimental games offer a means of studying fierce conflicts and hostile interactions without the usual ethical problems associated with the investigation of potentially antisocial forms of behaviour. Thirdly, the extremely abstract nature of the task confronting the subjects in a matrix game appears to enable "pure" strategic behaviour to be investigated in an idealized cognitive environment from which all irrelevant information and extraneous sources of variance have been removed. Lastly, experimental games are economical and easy to perform, yet they provide objective and quantitative information—which cannot be obtained in any other way— about important aspects of strategic decision making. In spite of this impressive catalogue of virtues, however, the experimental gaming tradition has been subjected to a number of searching criticisms, some of which are discussed later in this chapter.

In every finite, strictly competitive game, an optimal method of play can be prescribed for both players (see Chapter 4). Comparatively few experiments have been devoted to games of this type, most researchers preferring to concentrate on mixed-motive games of various kinds. There appear to be two main reasons for this state of affairs. The first is the belief of many researchers that the existence of rational solutions robs strictly competitive games of psychological interest: since the players' optimal strategies are prescribed by the minimax principle, experiments can reveal nothing more than the extent to which their choices are rational. Secondly, the fact that most everyday conflicts are mixed-motive is thought by many to render strictly competitive games less worthy of experimental investigation.

Both of these arguments can, however, be challenged. It should be borne in mind, first of all, that the minimax principle is not an entirely persuasive prescription for rational choice against an opponent who is expected to behave irrationally. Even in the simplest saddle-point game a non-minimax strategy may offer the best prospects against an opponent who is expected to deviate from minimax. In everyday life one is, of course, often confronted by opponents whose behaviour is unpredictable or systematically and predictably irrational. Against a rational opponent a minimax strategy is always optimal, but the solution is often far from obvious to someone unfamiliar with game theory and the minimax strategies may be difficult to find. It is clearly of some interest to investigate the factors influencing behaviour in such circumstances in order to discover some of the limits of human rationality. And although most everyday conflicts may indeed be mixed-motive, there is no doubt that many socially significant economic, military, political, and interpersonal conflicts are zero-sum (see Chapter 4); the rarity of strictly competitive conflicts has probably been greatly exaggerated (Kahan and Rapoport, 1974). For all of these reasons, the relative neglect of strictly competitive games by empirical researchers seems somewhat unwarranted.

In view of the fact that the literature on strictly competitive experimental games does not appear to have been reviewed in detail elsewhere, a fairly comprehensive survey is provided in this chapter. The survey is restricted chiefly to experiments on strictly competitive matrix games; the literature on two-person negotiation games has been thoroughly reviewed by Chertkoff and Esser (1976), Druckman (1977), Hamner (1977), Magenau and Pruitt (1979), Morley (1981), Morley and Stephenson (1977), Pruitt (1976), Rubin and Brown (1975), and Tysoe (1982) among others.

Section 5.2 contains a review of experiments based on games without saddle-points and Section 5.3 is devoted to experiments with saddle-point games. There is a certain amount of unavoidable overlap between these two sections since saddle-point and non-saddle-point games have occasionally been included in the same investigation. Section 5.4 contains a discussion of some of the major criticisms of experimental games, and in Section 5.5 a new experiment, designed to overcome some of these criticisms, is reported. A brief summary of the chapter is given in Section 5.6.

5.2 Review of Research on Non-Saddle-Point Games

Several early investigations centred on the behaviour of subjects competing in pairs in zero-sum games without saddle-points. Atkinson and Suppes (1958) and Suppes and Atkinson (1960, chap. 3) used three 2 × 2 games. Two were saddle-point games and in one of these both players had dominant strategies (see Section 4.5); the third was a non-saddle-point game. Each game was played 200 times in succession under conditions of incomplete information: the subjects were ignorant of the payoff matrices and were not directly informed about their opponents' choices. The results showed that minimax strategies "were not even crudely approximated" (Atkinson and Suppes, 1958, p. 374) in any of the games. In a later experiment by the same investigators (Suppes and Atkinson, 1960, chap. 9) subjects played a 2 × 2 non-saddle-point game 210 times in succession either under conditions of incomplete information or with a conventional payoff matrix. In neither case did the subjects conform at all closely to the minimax principle. Kaufman and Lamb (1967) investigated the behaviour of pairs of subjects over 100 repetitions of four 2 × 2 non-saddle-point games, and found that "under the conditions of the present experiment, players do not learn to play a game theory optimal strategy" (p. 958). The subjects in this experiment were given spinners resembling roulette wheels which could be used as randomizing devices to generate mixed strategies, but they made little use of them: "they [did] not play any mixed strategy in the sense of choosing a given alternative randomly with a fixed probability" (p. 959).

A few investigators have, however, reported a slight tendency for subjects to converge towards minimax strategies over trials when competing in pairs in free-play experiments. Sakaguchi (1960) examined the behaviour of two pairs of subjects in two simple saddle-point games, and found that in both cases one of the subjects approached the minimax strategy after 50 or 60 repetitions while the other did not. Malcolm and Lieberman (1965) reported a slight convergence towards minimax in the course of 200 free-play repetitions of a 2 × 2 non-saddle-point game: on the last 25 trials, 10 of the 18 subjects were conforming very roughly to the minimax prescription. Payne (1965) reported a similar small but significant tendency towards minimax in four 5 × 5 non-saddle-point games played 200 times in succession.

All of the experiments discussed so far involved free play between subjects competing in pairs, and for this reason the results are difficult to interpret. As explained in Section 4.4, a minimax strategy is not unambiguously "optimal" or "rational"—although it is by definition the safest course—if there is reason to believe that the opponent will deviate from minimax; in that case a non-minimax strategy may exploit the opponent's play to better advantage. Over repeated trials in an experiment, a subject may with good reason come to expect a non-minimax pattern of choices from his opponent, whereupon it is—arguably—rational for him to deviate from minimax himself in order to take maximum advantage of the opponent's "non-optimal" play. We shall return to this point later; for the present it is sufficient to point out that the interpretation of non-minimax strategies on the part of subjects in free-play experiments is highly problematical. For this reason a number of investigators have pitted subjects against computers or human opponents programmed to adopt minimax or non-minimax strategies, but unfortunately most have used 2 × 2 matrices, which leads to further problems of interpretation. The relevant findings are reviewed in the following paragraphs.

Lieberman (1960a, 1962) used a 2 × 2 game and reported a general convergence towards the "optimal" exploiting strategy on the part of subjects pitted against a non-minimax programmed opponent's strategy over 300 trials, but against a minimax program none of the subjects approached minimax at all closely. The failure of the subjects to respond to minimax with minimax has been interpreted, not only by Lieberman but also by subsequent commentators (see for example Rapoport and Orwant, 1962, p. 10) as indicating "irrational" or "non-optimal" play on the part of the subjects. But since a 2 × 2 game was used, this interpretation seems dubious: it was explained in Section 4.4 that all 2 × 2 non-saddle-point games have the property that if one player uses a minimax strategy, then the other player's expected payoff is the same *whatever* strategy he uses. Lieberman's subjects therefore lost nothing by using non-minimax strategies against the minimax program, and it seems unfair to interpret their behaviour in this treatment condition as irrational. In spite of this problem, virtually all subsequent investigations of responses to programmed strategies in non-saddle-point games have used 2 × 2 matrices, and non-minimax responses to minimax programs have almost invariably been interpreted as "non-optimal", "irrational", "incorrect" or "errors".

The following findings were also derived from investigations of responses to programmed strategies using 2 × 2 non-saddle-point games, and they are difficult to interpret for the reasons outlined above. Lacey and Pate (1960) reported that their subjects closely approximated minimax strategies against minimax programs and tended towards "optimal" exploiting strategies against non-minimax programs, especially when their opponents were non-human devices rather than human players. Kaufman and Becker (1961) used five 2 × 2 matrices, one of which had a saddle-point, and reported wild deviations from minimax in response to minimax programs in every game, especially in the saddle-point game. Pate (1967) used a 2 × 2 game whose value was negative from the subject's point of view; that is, the subject's expected payoff with optimal play on both sides was $-1/8$ per trial. No convergence towards minimax was observed over 120 trials, although the subjects who recognized that they were playing a losing game were less likely than the "non-recognizers" to accept the offer to play a further 120 trials. Pate and Broughton (1970) used a non-minimax program and found a slight tendency towards the "optimal" exploiting strategy over 240 trials. Fox (1972) found that his subjects tended to converge towards "optimal" exploiting strategies against a non-minimax program and towards minimax against a minimax program. The exploitation effect against a non-minimax program was replicated in an experiment by Fox amd Guyer (1973). Kahan and Goehring (1973) also reported an exploitation effect against a non-minimax program, and against a minimax program "subjects' performance was not optimal but was sufficiently and consistently close to it that arguments in terms of differences between perceived and objective probabilities provide an attractive explanation for the difference" (p. 27). Finally, Pate *et al.* (1974) replicated both the exploitation effect against a non-minimax program and the slight tendency towards minimax against a minimax program. They also reported that both effects were most pronounced in subjects who were relatively low in dogmatism.

The use of 2 × 2 matrices in all of the above experiments militates against a clear interpretation of the findings. The investigators evidently believe that subjects "ought" to use minimax strategies against minimax programs and "optimal" exploiting strategies against non-minimax programs if they are rational. But, in the first place, a non-minimax strategy is not unambiguously irrational or non-optimal against a minimax program in a 2 × 2 non-saddle-point game, since *any* pure or mixed strategy yields the same expected payoff in such a game. The fact that subjects tended to converge towards minimax against minimax programs in some experiments is somewhat surprising in view of the fact that they derived no advantage by doing so.

Secondly, against non-minimax programs, so-called "optimal" exploiting strategies are not unambiguously optimal. A deviation from minimax is clearly optimal only if the opponent's choices can be predicted with *certainty*; in the absence of certainty it is irrational according to game theory in so far as it exposes a player to the risk of outcomes *worse* than the worst that can result from a minimax strategy. In other words, against a non-minimax program a player has a perfectly rational justification—ensuring the best of the worst possible outcomes—for using a minimax rather than an "optimal exploiting" strategy.

In an unusually well designed experiment, Messick (1967) used a 3 × 3 non-saddle-point game, thus avoiding one of the major problems of interpretation mentioned above. Forty-two subjects played 150 repetitions of this game against a computer programmed to use either a minimax or one of two non-minimax strategies against them. Messick concluded from his results that "the study reported here unambiguously indicates that human Ss do not behave in a manner consistent with the minimax theory" (p. 46). In fact, Messick's sophisticated analysis of his data revealed no significant tendency among the subjects even to move towards minimax over trials in any of the treatment conditions; neither did they succeed in evolving "optimal" exploiting strategies against the non-minimax programs. These findings convincingly refute those of several less well designed experiments described above.

Before leaving strictly competitive non-saddle-point games, some brief mention is necessary of games of timing in pure conflict, particularly so-called silent duels. Games of timing are a class of two-person, zero-sum games in which each player's set of strategies is infinitely large and comprises the set of real numbers between zero and one. Unlike the more conventional matrix games, the problem facing the player is not *what* action to take, but rather *when* he should take action. The initial resources of the players are limited by the rules of the game and they are not necessarily equal, but each player can make only a fixed number of decisions to take action within the time interval $0 \leqslant t \leqslant 1$. At $t = 0$ every attempt fails, at $t = 1$ every attempt succeeds, and at any other time there is a positive probability of success. The typical Western duel between two gunfighters walking closer and closer towards each other is an example of a game of timing. Two classes of games of timing have been distinguished (Karlin, 1959). In the first class are so-called noisy duels, which are games of perfect information: when either player acts, his action and its effects are immediately known to his opponent. The second class comprises silent duels in which a player does not know when his opponent has fired and missed. In most discussions of games of timing to date, knowledge regarding a player's own and his opponent's initial resources is assumed given. Unlike many other infinite games, games of timing have minimax solutions analogous to those of finite games. Noisy duels have solutions in pure strategies—saddle-points—and silent duels have mixed-strategy solutions.

A detailed review of experiments concerning the behaviour of subjects in noisy and silent duels would be out of place in the present context, but a brief comment is desirable for the sake of completeness. In a review of several independent studies in this area, Rapoport, Kahan, and Stein (1976) report the mean correlation between observed firing times in 34 duels and the 34 prescribed minimax strategies to be 0.959. The authors interpret these findings to mean that the minimax principle is an extremely good predictor of the behaviour of subjects in games of timing, both in their saddle-point and non-saddle-point forms, although the latter appear to be somewhat more difficult for the subjects.

5.3   Review of Research on Saddle-Point Games

Very few empirical studies of behaviour in saddle-point games have been reported. One reason for this may be that the "correct" or "optimal" or "rational" strategies prescribed by formal game theory seem rather more obvious in these cases than in games requiring mixed strategies. The results do not, however, suggest that subjects invariably succeed in solving saddle-point games.

The experiment by Atkinson and Suppes (1958, Suppes and Atkinson, 1960, chap. 3), which incorporated both saddle-point and non-saddle-point games, has been discussed above, as has the mixed study of Kaufman and Becker (1961). Mention has also been made of several investigations of noisy duels, which may be regarded as infinite saddle-point games (Rapoport, Kahan, and Stein, 1976). Only a handful of other experiments on saddle-point games appear to have been published.

Lieberman (1959, 1962) used a very simple 2×2 saddle-point game repeated 200 times. The game was presented to the subjects in conventional matrix form. Of the 14 undergraduate subjects who participated in the experiment, 10 came to adopt the minimax pure strategy consistently and 4 exhibited behaviour similar but not identical to the prescribed pure minimax behaviour.

In a methodologically similar experiment with a slightly more complicated 3 × 3 saddle-point game (Lieberman, 1960b), the following results emerged. About half of the 30 undergraduate subjects conformed to the minimax prescription 100 per cent of the time after between 10 and 125 trials, and on the final 20 trials 94 per cent of the subjects were making consistent minimax choices. Many of the non-minimax choices made by the subjects in the later trials of this experiment were, on their own accounts, motivated by a desire to alleviate the boredom of the task.

Morin (1960) presented 28 undergraduate subjects with 28 different 2 × 3 games, all of which had saddle-points. In addition, the minimax strategy could be found in each case by eliminating dominated strategies from both players' repertoires. Each subject, acting as the player with two pure strategies, made one choice on each of the 28 matrices. Out of 784 choices, 140 were "errors" (non-minimax choices). Morin noted that "errors" were made more often in games in which the average expected value (assuming random behaviour on the part of the opponent) of the non-minimax strategy was greater than that of the minimax strategy.

Suppes and Atkinson (1960, chap. 6) reported an experiment involving two 2 × 4 saddle-point games. The subjects (96 undergraduate subjects) were randomly assigned to one of two treatment conditions in which they competed in pairs; the two conditions involved slightly different payoffs. In each case 200 repetitions of the game were played. The experiment was somewhat unusual (although similar to an experiment by the same investigators on non-saddle-point games described above) in that the subjects were not shown the payoff matrix and had to infer it as the experiment proceeded. They were, however, aware of each other's existence, and of the nature of the task. The results were summarized as follows: "The observed asymptotic response probabilities . . . clearly show that the pure game-theory strategies are not even roughly approximated. There is not even any appreciable tendency for the observed probabilities to move away from the learning-theory predictions and toward the optimal game-theory strategies" (p. 151). In interpreting these results it should, however, be borne in mind that, on account of the subjects' ignorance of the payoff matrices, and other peculiarities in the design, this experiment was not concerned with choices under complete information typical of other investigations in this area.

Brayer (1964) presented 100 undergraduate subjects with 90 3 × 3 games, all of which had saddle-points. Subjects played three trials on each game against an opponent who either chose randomly or else adhered consistently to the minimax strategy. Against the randomizing opponent, a subject's highest expected value was attainable in each case by consistently choosing one of his non-minimax pure strategies. The results showed that subjects adopted their minimax strategies 0.59 times on average (out of a possible maximum of 3) when playing against a randomizing opponent, compared with an average of 2.75 against a minimax opponent. A subsidiary finding of interest was that subjects used the minimax strategy more often when the absolute value of the game, which represented the best payoff they could guarantee themselves on each trial, was high than when it was low, although according to formal game theory the absolute value is irrelevant to the logic of the situation.

The findings on strictly competitive saddle-point games are not quite as confusing and confused as those on non-saddle-point games, if only because there are fewer of them. The problems associated with free-play experiments and the interpretation of non-minimax choices in response to non-minimax opponents' strategies bedevil several of the experiments reviewed in this section and it is unnecessary to reiterate the comments about these matters made in Section 5.2. In spite of these problems, certain general conclusions seem warranted.

In all but two of the studies devoted to saddle-point games, wild departures from minimax were observed in the subjects' strategy choices. In the remaining two studies, Lieberman (1959, 1962) found that most subjects managed to converge on the minimax strategy after numerous trials, and Brayer (1964) reported a much stronger predilection for minimax among subjects exposed to a minimax opponent's strategy than among those exposed to a non-minimax opponent's strategy. This latter finding highlights the weakness of the minimax principle in circumstances in which an opponent is expected to behave "irrationally" since it was arguably in the subjects' rational self-interest in one of Brayer's treatment conditions to deviate from the minimax principle, and that is precisely what they tended to do.

5.4   Critique of Experimental Gaming

The experiments reviewed in Section 5.2 and 5.3 represent only a tiny fraction of the experimental gaming literature, most of which is devoted to mixed-motive games rather than to the strictly competitive variety. In spite of its popularity, however, the whole research tradition has been subjected to a number of criticisms which are worth examining in some detail.

Over the years, a growing chorus of commentators have expressed misgivings about the ecological validity of experimental games. Ecological validity, which became a fundamental concern of social psychologists during the 1970s, is the extent to which the results of an experiment can be generalized to non-experimental, naturally occurring situations. It should be carefully distinguished from internal validity (the extent to which the conclusions of an experiment are true within the limits of the particular subjects and methods used) and external validity (the extent to which the conclusions are generalizable beyond the experimental subjects and methods used). Experimental games, because of their abstract and highly artificial nature, seem especially vulnerable to criticisms regarding their ecological validity. A typical comment along these lines is the following:

> A price must be paid in moving from complex situations in the world to simple games in the laboratory. When people are put in a simple artificial situation, one might argue, they will behave in ways appropriate to a simple artificial situation, thereby revealing nothing about how they will behave in a complex real situation (Hamburger, 1979, p. 231).

One of the strongest critics has been Charlan Nemeth, who has argued that the apparently irrational behaviour of subjects in experimental games "is due primarily to the essential incomprehensibility of the situation in which the subject is placed" (1972, p. 213).

Researchers have responded to this line of criticism in two different ways. Some have claimed that behaviour in abstract laboratory games is sufficiently interesting and important in itself to warrant continued research without making any claims about behaviour in naturalistic interactions. The most influential spokesman for this point of view has been Anatol Rapoport, a uniquely important figure in the experimental gaming field on account of his numerous research reports, books, and review articles, and above all his editorship during the late 1960s and 1970s of the gaming section of the *Journal of Conflict Resolution*. Rapoport has consistently argued against attempts to generalize the findings of experimental games beyond the abstract and idealized laboratory situations on the grounds that the same laws do not govern "both the events in the laboratory and those of the cosmos" (1970, p. 40). Most researchers, on the other hand, are in favour of generalizing from the laboratory to the cosmos in spite of the technical and theoretical problems which this entails. Pruitt and Kimmel (1977), for example, believe that "there is continuity between the laboratory and the real world" (p. 367) and that "it is preferable for researchers to try to generalize their findings, because an analysis of limitations to plausible generalization can stimulate hypothesis building and the development of new research tasks" (p. 368).

One cannot help being struck, however, by the paucity of "new research tasks" that have sprung from concern with the ecological validity of experimental games. In particular, few attempts have been made to compare behaviour in abstract experimental games with behaviour in more lifelike strategic interactions. After a comprehensive survey of research on mixed-motive games, Wrightsman, O'Connor, and Baker (1972) had this to say:

> What surprises us most, in our review of research, is that apparently no studies have compared [behaviour] in a laboratory . . . game with

[behaviour] in different real-world tasks. While artificiality can also be assessed through laboratory manipulations, comparisons of . . . behaviour across settings should be undertaken (p. 277).

Tedeschi, Schlenker, and Bonoma (1973) have stated bluntly that "criteria for the assessment of ecological validity are nonexistent for experimental games" (p. 202) and that "no generalizations about social phenomena based exclusively on laboratory experiments could be safely assumed to be applicable in natural social environments without further inquiry" (p. 203). In that case, one is tempted to retort, let us engage forthwith in the necessary "further inquiry"! Schlenker and Bonoma (1978) suggested in a later publication that "theoretical considerations determine judgments about generalizability, and boundary experiments can be performed to assess the limits of the hypotheses" (p. 33). This challenge has not been generally heeded, but a "boundary experiment" designed to provide "criteria for the assessment of ecological validity" is reported in Section 5.5, and others are reported in subsequent chapters of this book.

A second major criticism—at least as serious as the challenge regarding ecological validity, though less often voiced—concerns the uncertain payoff structures of experimental games. The experiments purport to investigate behaviour in games with precisely specified strategic properties, and the assumption is always made that the subjects' preferences among the outcomes correspond exactly to the various points, tokens, or small monetary rewards assigned to them by the experimenter. But there is every reason to believe that the subjects' preferences may be influenced by factors outside the explicit payoff structures of the games. It is not difficult to imagine circumstances in which an experimental subject may, for example, prefer to receive the same payoffs as his opponent rather than to maximize his own gain at his opponent's expense. This implies that an investigator's assumptions about the payoff structure of an experimental game may not accurately reflect the utilities governing the subjects' choices.

The problem has been expressed most forcefully by Erika Apfelbaum (1974):

> When used in social psychological studies, however, the matrix is a payoff device; it does not refer to utilities or, speaking more loosely, to the subjective values of the different outcomes. These values are initially unknown. . . . There is no reason to assume that they are fixed *a priori* (p. 108).

This criticism is potentially crippling to the whole experimental gaming tradition, implying as it does that investigators are simply ignorant of the underlying strategic structures of the games which their subjects have played. There is no guarantee that experiments purporting to investigate behaviour in specific games do not reflect behaviour in games with quite different—and unknown—strategic structures.

Unless a close correspondence can be demonstrated between the subjective preferences of the subjects and the explicit payoffs assumed by the experimenter, the results of an experimental investigation are open to serious doubt. What are urgently needed are studies in which this problem is examined empirically. The experiment reported in Section 5.5, and the other new experiments reported later in this book, incorporated direct measures of the subjects' preferences in order to discover how closely they corresponded to the explicit payoff structures of the games which the subjects were assumed to be playing.

5.5   Experiment I: Abstract and Lifelike Strictly Competitive Games

The experiment described in this section (Colman, 1979a, Experiment I) represents a radical departure from traditional research methods. Like the other new experiments reported in this book, it was designed to take account of the major criticisms of experimental gaming discussed in Section 5.4. The experimental design enabled a direct comparison to be made between behaviour in two structurally equivalent but psychologically different decision contexts: (a) an abstract matrix game similar to those used in previous research on strictly competitive experimental games, and (b) a simulation of a lifelike strategic interaction having an identical payoff structure to the matrix. Secondly, a special design feature permitted an investigation to be made of the degree to which the subjects' preferences corresponded to the explicit payoff structure of the game.

The simplest type of strictly competitive payoff structure was used, namely a 2 × 2 saddle-point game. In the light of previous findings reviewed in Section 5.3, the payoff structure was constructed in a manner calculated to elicit a substantial proportion of non-minimax choices from the subjects. First, following Brayer's (1964) finding, the value of the game was negative from the subjects' point of view; in other words it was a "losing game". Secondly, in view of Morin's (1960) results, the expected payoff of the subjects' non-minimax strategy—assuming that the opponent randomizes equally between his available options—was higher than that of their minimax strategy. The subjects made 30 choices against programmed opponents' strategies that were either consistently minimax, consistently non-minimax, or random with equal probability assigned to each option.

It was hypothesized that significant differences would be observed between behaviour in the abstract and lifelike decision contexts. In particular, fewer minimax choices in response to the consistent minimax opponent's strategy were anticipated in the lifelike simulation than in the abstract matrix game. This hypothesis was based on the common-sense assumption that superfluous information is likely to distract attention from the essential strategic ingredients of the situation: it is especially difficult to make coldly rational choices in a "rich" decision context filled with irrelevant facts.

A second hypothesis was that the frequency of minimax choices would be greatest in response to the consistent minimax opponent's strategy and least in response to the consistent non-minimax opponent's strategy, and that this difference would become increasingly pronounced over trials. This second hypothesis was based on previous research (e.g. Brayer, 1964) indicating that subjects tend to meet minimax with minimax and non-minimax with exploiting strategies calculated to take advantage of the opponent's irrational behaviour in spite of the attendant risks. Both the consistent and random non-minimax opponents' strategies could be exploited in this experiment by non-minimax counter-strategies, but it was assumed that the subjects would be able to infer the opponent's consistent non-minimax strategy more quickly and confidently than the random opponent's strategy in the course of repeated trials.

Method

In the abstract decision context, pairs of subjects were presented with Matrix 5.1. The matrix, which has a saddle-point in the bottom left-hand corner, was explained to the subjects at great length, and they were quizzed by the experimenter until it was clear that they understood it fully. They were then presented with the following typewritten instructions:

Matrix 5.1

		Opponent	
		L	R
Subject	L	−2	2
	R	−1	0

Your task will consist of making a series of 30 decisions. You may each earn up to 120 points if your decisions turn out well, but you may get less. How many points you eventually end up with will depend not only on the decisions which you make but also on the decisions which the other person makes. You are advised, therefore, to consider each decision carefully.

After each joint decision you will receive −2, −1, 0, 1, or 2 points. Your decision, as shown on the cards in front of you, will in each case be either R for Right or L for Left. The scheme for awarding points is summarized on the payoff diagram. . . . You can tell which colour applies to you by the colour of the cards in front of you. This will also tell you whether you are choosing between the rows marked L and R (that's if you are Blue) or the columns marked L and R (if you are Green).

By examining the payoff diagram you can easily see what the outcome of each joint decision will be for each of you. If, for example, Blue and Green both choose L, then Blue gets -2 points and Green gets 2. If Blue and Green both choose R, then they each get 0 points. If Blue chooses L and Green chooses R, then Blue gets 2 points and Green gets -2. Finally, if Blue chooses R and Green chooses L, then Blue gets -1 point and Green gets 1 point.

Subjects in the simulation conditions were given no payoff matrix, but were presented with the following version of the game in typewritten form:

Your task will consist of making a series of 30 decisions. You may do very well for yourself if your decisions turn out well, but you may be less successful. The outcome in each case will depend not only on the decisions which you make, but also on the decisions which the other person makes. You are advised, therefore, to consider each decision carefully.

Your decisions will be based on the following hypothetical situation. You and an opponent from a rival party are the only candidates running for election in a particular constituency. At frequent intervals the local radio station invites both of you to participate in a live debate against each other in order to help the listeners to make up their minds. The problem facing you on each occasion is whether to agree to take part in such a debate or whether to refuse the invitation, and your sole consideration is to get as many votes as possible. In each case your decision, as shown on the cards in front of you, will be AGREE or REFUSE.

The following facts were told to you by a public opinion research organization which programmed a computer with all the relevant information about the election (including the candidates' public images) and these facts you know to be perfectly accurate. If neither candidate agrees to the debate, the radio station will not bother to publicize the fact, so naturally neither candidate will lose any votes to his opponent. If you refuse to debate and your opponent agrees, however, this fact will be made known to the public, and a significant number of your supporters will transfer their allegiance to your opponent. If, on the other hand, you both agree to the debate, you will inevitably be outshone in the minds of some listeners and you will lose twice as many votes to your opponent as you would have lost in the event of your refusing to take part while the opponent agreed to. In the event of your agreeing to the debate and your opponent refusing, however, you will gain as many votes from your opponent as you would have lost in the event of the debate actually taking place and your image thereby damaged.

Summary:—

You refuse, he refuses: you lose/gain 0 votes, he loses/gains 0 votes;

You refuse, he agrees: you lose x votes, he gains x votes;

You agree, he refuses: you gain $2x$ votes, he loses $2x$ votes;

You agree, he agrees: you lose $2x$ votes, he gains $2x$ votes.

The subjects, 42 male and 42 female undergraduate and postgraduate students at the University of Leicester, were randomly assigned to treatment conditions. They were tested in pairs in order to create the impression that they were interacting with each other, although in reality each subject was exposed to one of the programmed opponent's strategies described earlier. After each trial the alleged outcome was communicated to each subject, and each was requested to indicate his degree of satisfaction with the outcome on a five-point rating scale from "very displeased" to "very pleased". Each subject's satisfaction ratings were later to be compared with his corresponding payoffs (points or votes) in order to gauge the agreement between subjective preferences and explicit payoffs.

After 30 trials the experimenter interviewed each subject separately. The post-experimental interview included the following questions: "How do you feel about the overall results?"; "How do you feel about your partner?"; "What was your general strategy?"; and "Do you have any further comments?". The subjects were then debriefed and thanked for their participation.

Results

The main dependent variable was the number of minimax choices made by the subjects. The results were tabulated in three trial blocks of 10 choices each. The mean number of minimax choices per trial block across all treatment conditions was 3.98 (out of a possible total of 10), indicating that the subjects departed frequently from minimax. This should be interpreted in the light of the fact that two-thirds of the subjects were exposed to non-minimax opponent's strategies against which non-minimax choices are (arguably) optimal.

A three-way Analysis of Variance (Decision Context × Opponent's Strategy × Trial Blocks), supplemented by *a posteriori* Tukey tests, revealed significant effects due to all three independent variables.

The matrix elicited significantly more minimax choices, with a mean of 4.88 per trial block, than the lifelike simulation in which the mean was 3.07. This was due very largely to the fact that subjects in the matrix decision context increased the frequency of their minimax choices over trial blocks in response to the consistent minimax opponent's strategy while subjects in the lifelike simulation failed to do so. This finding, which confirms one of the major hypotheses, implies that only in the matrix game did the subjects learn the prudence of minimax in response to minimax through experience in the game. The frequency of minimax choices in response to the consistent non-minimax opponent's strategy declined over trial blocks in both the matrix and the lifelike decision contexts as expected, but only in the matrix game was a decline observed in response to the random opponent's strategy. Averaged over trial blocks, significantly more minimax choices were made in response to the consistent minimax than to the random, and in response to the random than to the consistent non-minimax opponent's strategy, which confirms the second major hypothesis. These differences were, however, much more pronounced in the matrix than in the lifelike decision context.

For each subject, a product–moment correlation coefficient was computed between his payoffs (points or votes) on each trial and his satisfaction ratings of the corresponding outcomes. A perfect positive correlation of 1.00 suggests that the subject was motivated solely by the explicit payoffs, while lower correlations indicate the importation of extraneous sources of utility and departures from the explicit payoff structure. The correlations were positive for 82 of the 84 subjects and the overall grand mean was 0.61, which suggests a strong but by no means perfect correspondence between explicit payoffs and subjective preferences.

An Analysis of Variance of the payoff–satisfaction correlations revealed a number of interesting differences. The correlations were slightly (though significantly) higher in the matrix decision context, with a mean of 0.62, than in the lifelike simulation in which the mean was 0.59. The rich context of information surrounding the lifelike simulation evidently provided subjects with more scope for introducing extraneous sources of satisfaction and dissatisfaction than did the relatively austere matrix decision context, but what is most surprising is how small the difference turned out to be.

The opponent's strategy also had a significant effect on the correlations: the highest mean correlation (0.84) was found in response to the random opponent's strategy, followed by the consistent non-minimax (0.65) and the consistent minimax opponent's strategies (0.33), and all three differences between these means were significant. The explanation is probably twofold: it is when an opponent is behaving predictably (consistent minimax or consistent non-minimax) that a subject is most likely to contemplate on extraneous factors in the situation, and when a subject is repeatedly losing points or votes (against the consistent minimax opponent's strategy) he may be most strongly tempted to seek sources of satisfaction outside the explicit payoff structure.

Since many of the subjects evidently departed from the explicit structure of the game in spite of admonitions to base their choices solely on the "official" payoffs, the variations in choice behaviour across treatment conditions are difficult to interpret. It is conceivable that subjects in different treatment conditions were in reality playing games with different underlying strategic structures, and this may account for the significant differences in choice behaviour. In order to examine this possibility, a supplementary statistical analysis was performed on the choices of the subjects who manifested the highest payoff–satisfaction correlations, that is, those who adhered extremely closely to the explicit payoff structure of the game. The choices of 18 subjects, nearly all of whom had produced correlations in excess of 0.90, were used in the supplementary analysis. All of the original findings were replicated and the pattern of results was strikingly similar, thus indicating that the major findings were not artifacts.

The results of the post-experimental interviews suggest that, on the whole, the subjects found the tasks ego-involving and interesting. A few (12 of the 84) described the task as "rigged" or "fixed" or suggested that there was "no real opponent" or that "incorrect feedback was given". Comments of this type were not made by any of the subjects exposed to the consistent minimax opponent's strategy; it was only when the opponent deviated from the strategy prescribed by formal game theory (consistently choosing the non-minimax option or randomizing) that a minority of the subject became suspicious.

The subjects in this experiment played a losing game whose value was negative from their point of view. Eight subjects described the game in the post-experimental interview as "unfair", "biased", "one-sided", and so on, but comments in this category were not made by any of the subjects in the lifelike simulation. Strategic interactions in everyday life are, of course, often unfair, and the subjects in the lifelike simulation did not consider the negative value of the game to be worthy of comment. But the subjects in the matrix game who commented in this way were expressing an implicit assumption that an abstract, game-like task ought to be fair. This assumption was probably imported from experience in recreational games which are normally fair in the sense of having zero values. If this interpretation is correct, it suggests a tendency on the part of some subjects to interpret a matrix game as a parlour game.

Many of the subjects revealed an intuitive appreciation of the rationality of the minimax principle in their comments about their (alleged) opponents. Twenty subjects described their opponents in terms like "clever", "sensible", "did as I would have done", and the like, but none of the subjects exposed to the consistent non-minimax opponent's strategy made comments of this type, and only three exposed to the random opponent's strategy did so. This pattern was reversed for comments to the effect that the opponent was "silly", "stupid", "lacking in understanding", and so on; of the 25 subjects who commented in this way, all but two were exposed to one of the non-minimax opponent's strategies.

Discussion

When confronted with an opponent's strategy that was rational according to the dictates of game theory, the subjects in the conventional matrix game behaved more rationally than those in the lifelike simulation. Against an irrationally randomizing opponent's strategy, the subjects in the matrix game were also more successful in exploiting the situation to their own advantage. The consistent minimax opponent's strategy elicited the most frequent minimax choices from the subjects, and the consistent non-minimax opponent's strategy elicited the fewest, and this difference was more pronounced in the matrix game than in the lifelike simulation. The findings are in harmony with some previously reported results (e.g. Brayer, 1964), but they suggest that extrapolations from abstract matrix games to strategic interactions in natural environments may be unwise.

The simulation used in this experiment was, of course, unrealistic and artificial to a degree. The subjects were required to imagine themselves in a predicament that they were not really in, they were barred from communicating with their alleged adversaries, and they made their choices rather more rapidly than they would have done in the natural course of events. Above all, they were in possession of complete information regarding the outcomes of all possible combinations of strategies, a condition which is seldom met in everyday life. The simulation nevertheless represented a more realistic version of the game than the traditional payoff matrix. The lifelike simulation occupies an intermediate position between a meaningless, abstract matrix and a naturally occurring strategic interaction. It is not known whether behaviour in the lifelike simulation accurately reflects the way the subjects would have behaved had the imagined situation actually arisen, but it is almost certainly a better reflection of their likely behaviour than the data from the matrix game.

An important feature of the experiment was the attempt to gauge the degree to which subjects imported extraneous utilities into the game, thereby effectively altering its strategic structure. The evidence showed that many of the subjects departed from the explicit payoff structure in this way. Deviations were barely greater in the lifelike simulation than in the matrix game. One of the major attractions of matrix games, from the researcher's point of view, is the accurate control which can supposedly be exercised over the nature of the subjects' strategic interdependence. The results of this investigation suggest, however, that this control is far from perfect in a matrix game, and that it is hardly worse—at least in this case—in a lifelike simulation. A different lifelike simulation with the same payoff structure might, of course, have produced quite different results, but the findings of this experiment do not lead one to expect that adherence to explicit payoffs is necessarily maximized by using abstract matrices. Most importantly, the results have shown that it is both feasible and desirable to monitor subjective preferences in experimental games.

5.6 Summary

The relationship between game theory and experimental games was discussed in Section 5.1. Game theory cannot be empirically tested and falsified on account of its non-predictive character, but experimental games have the potential to reveal how fallible human decision makers behave in strategic interactions whose structural properties are understood in terms of the theory. The experimental literature on strictly competitive games was comprehensively reviewed in Sections 5.2 and 5.3, and it was argued that very few clear-cut conclusions are justified on the basis of the published research. Section 5.4 was devoted to a critique of experimental gaming in general, particularly on the grounds of (a) the doubtful ecological validity of conventional experiments, and (b) the uncertain nature of the subjects' preferences among the outcomes in laboratory games. In Section 5.5 a new experiment designed to overcome these criticisms was reported. Behaviour in a matrix game was shown to differ in certain important ways from behaviour in a structurally equivalent lifelike simulation of a strategic interaction, and the preferences of many subjects failed to correspond to the explicit payoff structure of the game in both the matrix and the lifelike simulation. Many subjects showed an intuitive appreciation of the minimax principle but deliberately deviated from it in order to maximize their payoffs against "non-optimal" opponents' strategies; this finding highlights the weakness of the minimax principle as a prescription for rational choice against an irrational opponent.

6

Two-Person, Mixed-Motive Games: Informal Game Theory

6.1 Mixed-Motive Games

GAMES in which the players' preferences among the outcomes are neither identical (as in pure coordination games) nor diametrically opposed (zero-sum) are called mixed-motive games. This term draws attention to the complex strategic properties that motivate the players partly to cooperate and partly to compete with one another. A player in a mixed-motive game has to contend with an *intra*personal, psychological conflict arising from this clash of motives in addition to the *inter*personal conflict that exists in the game.

At an abstract level, a mixed-motive game can be distinguished from a zero-sum game by the fact that the sum of the payoffs differs from one outcome to another; it is not the case that what one player gains the other(s) must necessarily lose and vice versa. Mixed-motive games are therefore sometimes called *variable-sum* or *non-zero-sum* games, but this can lead to misunderstandings in view of the fact that pure coordination games are also variable-sum.

Many of the concepts of formal game theory introduced in Chapter 4 can be applied to mixed-motive games, but purely logical solutions are generally not forthcoming and in practice it is necessary to resort to informal or—what amounts to the same thing—common-sense analyses of individual cases in order to discover sensible methods of play. The mathematical inconclusiveness of mixed-motive games is, however, balanced by an abundance of psychologically interesting phenomena for which they provide precise models, such as cooperation and competition, risk taking and caution, trust and suspicion, altruism and spite, threats, retaliations, and commitments.

This chapter is concerned with two-person mixed-motive games; multi-person games are discussed in Chapter 8. The following section provides a framework for classifying the simplest kinds of mixed-motive games. Sections 6.3, 6.4, 6.5, and 6.6 contain outlines of the formal and informal properties of the four most interesting examples of such games, and in Section 6.7

the similarities and differences between them are discussed. Section 6.8 is devoted to an outline of metagame theory and of a possible solution to the Prisoner's Dilemma game. A brief summary of the chapter is given in Section 6.9.

6.2   Classification of 2 × 2 Mixed-Motive Games

The range of strategic possibilities in mixed-motive interactions is so unimaginably vast that theorists and experimentalists have tended to concentrate mainly on the simplest cases. Since a strategic interaction requires a minimum of two decision makers each faced with a choice between a minimum of two options, the simplest conceivable mixed-motive games are 2 × 2.

In formal game theory, the payoffs are assumed to be measured on interval scales (see Sections 1.2 and 4.7) and an interval-scale classification scheme for all 2 × 2 games has been devised by Richard Harris (1969, 1972). For most practical purposes, however, a simpler ordinal-scale scheme proposed by Rapoport and Guyer (1966) is sufficient. This system ignores differences of degree and considers only the order of preference of the outcomes, from best to worst, from the point of view of each player. Games in which the ordinal relations among the payoffs are the same are regarded as strategically equivalent in this system of classification. Under these bold assumptions, Rapoport and Guyer were able to show that there are exactly 78 strategically distinct 2 × 2 games. (No-one has yet attempted to classify larger games in this way because there are, for example, in excess of 1828 million strategically distinct 3 × 3 games!)

Of the 78 2 × 2 games, 12 are symmetric in the sense that the situation is unchanged if the players swap roles. Eight of these symmetric games possess *optimal equilibrium points* corresponding to dominant strategies of both players, and these are considered to be theoretically uninteresting from a strategic point of view. An example of such a game is shown in Matrix 6.1.

Matrix 6.1
Game with an optimal equilibrium point

	II	
	1	2
1	4,4	2,3
2	3,2	1,1

(I on the left, II on top)

It is customary to display both players' payoffs in the matrix because it cannot be assumed—as in a zero-sum game—that Player II's payoffs are simply the negatives of Player I's. The first number in each cell is Player I's payoff and the second is Player II's. These numbers, of course, have only ordinal significance in the present context, that is, 4 is "best", 3 is "second best", and so on. It is clear that both players have dominant strategies and that the corresponding outcome is optimal for both. Row 1 is better from Player I's point of view than row 2 whichever column Player II may choose, and column 1 is better for Player II than column 2 against either of Player I's options. If both players choose their dominant strategies—clearly the only rational way to play this game—then both receive their best possible payoffs. The outcome in the top left-hand cell is an optimal equilibrium point because neither player has any incentive to deviate from it. Experimental evidence outlined in Chapter 7 reveals, however, that subjects do not invariably choose their dominant strategies in games with optimal equilibrium points. This suggests that their subjective preferences do not correspond to the explicit payoff structures of the games they are supposed to be playing. A player may, for example, prefer to spite the "opponent" by maximizing the difference between their respective payoffs rather than simply to aim for the best payoff from his own viewpoint. This implies, of course, that the player has changed the payoff structure of the game.

The four symmetric 2 × 2 games without optimal equilibria are strategically interesting, and two of them have attracted a great deal of interest among experimental investigators. The four basic structures have been described by Rapoport (1967b) as the archetypes of the 2 × 2 game. The least well known of these archetypes is discussed in the following section.

6.3   Leader

The payoff structure of the game of Leader is shown in Matrix 6.2. This strategic structure arises in numerous everyday interactions, but the following familiar example will suffice to illustrate its essential properties. Two motorists are waiting to enter a heavy stream of traffic from opposite ends of an intersection, and both are in a hurry to get to their destinations. When a gap in the traffic occurs, each must decide whether to concede the right of way to the other (C) or to drive into the gap (D). If both concede, both will be delayed, which is the second-to-worst outcome for both (2,2). If both drive out together, a collision may occur (1,1). But if one drives out while the other concedes the right of way, then the "leader" will be able to proceed on his journey immediately and the "follower" may be able to squeeze in behind him before the gap closes (4,3 or 3,4). The payoffs shown in Matrix 6.2 agree with common-sense assumptions about the motorists' orders of preference among the possible outcomes.

Matrix 6.2
Leader

II

		C	D
I	C	2,2	3,4
	D	4,3	1,1

It is clear that there are no dominant strategies in the game of Leader; neither player has a strategy that yields a preferable payoff irrespective of the other's choice. According to the minimax principle—the cautious policy of choosing so as to avoid the worst possible outcomes—both players should choose C; this ensures that neither can receive a payoff less than 2 (the possibility of a collision is ruled out). But the minimax strategies are not in equilibrium. Each player would have cause to regret his choice when the other's strategy was revealed. This fact illustrates the complete collapse of the minimax principle as a prescription for rational choice in mixed-motive games.

There are in fact two pure-strategy equilibrium points in Matrix 6.2, in the top right-hand and bottom left-hand cells respectively. If Player I chooses D, II can do no better than to choose C and vice versa; neither can benefit by deviating unilaterally from such an equilibrium outcome. But whereas in strictly competitive games equilibrium points are always equivalent, in Leader Player I prefers the D,C equilibrium and Player II the C,D one. Formal game theory provides no satisfactory way of resolving this difference of opinion without explicit negotiation. In a lifelike strategic interaction with the structure of Matrix 6.2, however, a sensible way of choosing may emerge from the fact that one of the equilibrium points is more *prominent* to both participants than the other. The concept of prominence was discussed at length in Chapter 3 in connection with pure coordination games. In the example above of manoeuvring in traffic, cultural factors outside the abstract structure of the game may guide the motorists towards a practical solution based on prominence. They may, for example, tacitly agree upon the principle of "first come, first served", or if one is a (chivalrous) man and the other a (non-feminist) woman, on the principle of "ladies first", or if the game is repeated a number of times the players may agree tacitly to alternate in playing the roles of "leader" and "follower". In sharp contrast to strictly competitive games, it can be in the players' interests in a mixed-motive game to communicate their intentions to each other; the motorists may gesticulate or flash their headlamps, for example. Informal considerations like these are often necessary in order to find sensible ways of playing mixed-motive games which, after all, share some of the properties of pure coordination games. But there is no completely rigorous criterion of rationality in these cases.

6.4   Battle of the Sexes

The second of the four archetypal 2 × 2 games is depicted in Matrix 6.3. The following well-known predicament, christened "Battle of the Sexes" by Luce and Raiffa (1957, p. 91), illustrates the payoff structure of Matrix 6.3. A married couple has to choose between two options for an evening's entertainment. The man prefers one kind of entertainment and the woman the other, but both would rather go out together than alone. If both opt for their first choices (C, C), each ends up going out alone (2,2), and a worse outcome (1,1) results if both make the heroic sacrifice of going to the entertainments they dislike (D, D). If one chooses his or her preferred option and the other plays the role of "hero", however, the outcome (4,3 or 3,4) is better for both, but less so for the "hero". Luce and Raiffa's (pp. 90–1) abstract model of this marital problem is not quite identical—and less intuitively persuasive—than the one shown in Matrix 6.3.

Matrix 6.3
Battle of the Sexes

		II	
		C	D
I	C	2,2	4,3
	D	3,4	1,1

Battle of the Sexes resembles Leader in many ways. Neither player has a dominant strategy, the minimax strategies intersect in the non-equilibrium (C, C) outcome, and both C, D and D, C are in equilibrium. A player who deviates unilaterally from minimax in Battle of the Sexes, however, rewards the other player more than himself and may therefore be described as a "hero" (Rapoport, 1967b), whereas a player who deviates unilaterally in Leader rewards himself more than the other. In Battle of the Sexes, as in Leader, a player can gain an advantage by communicating with the other player in order to indicate a binding commitment to the D strategy. In the marital example the man may, for instance, announce that he is irrevocably committed to his chosen entertainment. If the woman acts in her own best interests, this will work to his advantage, but he has the problem of convincing her that his commitment is really binding. One way of solving this problem would be to produce a ticket he has already bought although, as Luce and Raiffa (1957) have pointed out, "to some spirited females, such an offhand dictatorial procedure is resented with sufficient ferocity to alter drastically the utilities involved in the payoff matrix" (p. 91).

The problem of commitment plays a central role in many mixed-motive interactions. I have commented on it elsewhere:

> The most striking conclusion which has emerged from a theoretical analysis of such situations is somewhat counterintuitive; it contradicts conventional wisdom and "common sense". It is that bargaining power derives largely from some voluntary and irreversible sacrifice of freedom of choice. "Keeping all options open" is a faulty policy based on incomplete understanding, and one which may significantly weaken one's bargaining position. The practice among invading armies of burning their bridges (or boats) and thereby voluntarily relinquishing the option of withdrawing is strategically analogous to the tactic of the suffragettes of chaining themselves to the railings. By this action they rendered themselves impervious to threats to disperse or be punished (Colman, 1975, p. 20).

The strategy of commitment arises in an especially ugly form in the following game.

6.5   Chicken

The third of the archetypal 2 × 2 games, shown in Matrix 6.4, is one that has attracted a certain amount of empirical research.

Matrix 6.4
Chicken

		II	
		C	D
I	C	3,3	2,4
	D	4,2	1,1

Matrix 6.4 depicts the ordinal structure of Chicken, the prototype of the "dangerous game" (Swingle, 1970a). Its name derives from various versions of a gruesome pastime that originated among Californian teenagers in the 1930s. (The game itself is, of course, older than that. Schelling, 1966, p. 117, cites an example from Homer's *Iliad* in which Antilochos won a game of Chicken against Menelaos 3000 years ago.) The most familiar version is this: two motorists speed towards each other along a narrow road. Each has the

option of swerving to avoid a head-on collision and thereby showing himself to be "chicken" (*C*) or of resolutely driving straight ahead (*D*). If both players are "chicken", the outcome is a draw (3,3) and if both drive straight ahead, they risk death or serious injury (1,1). But if one "chickens out" by swerving while the other exploits his adversary's caution by driving straight on, the "chicken" loses face (though not his life) and the "exploiter" wins a prestige victory on account of his boldness or *machismo* (2,4 or 4,2).

Once again there are no dominant strategies and the minimax strategies intersect in the *C, C* outcome. This outcome is not an equilibrium point, so a player can benefit by deviating from it while the other sticks to minimax. The equilibrium points in Chicken are the asymmetric, *C, D* and *D, C* outcomes. The "exploiter" who deviates unilaterally from minimax gains an advantage and, in contrast to the games examined above, invariably affects the other player adversely by so doing. By trying to get the maximum payoff for himself, the "exploiter" not only harms the other player but also exposes himself and the other to the risk of a disastrous outcome. This is what makes it a dangerous game.

The game of Chicken possesses a number of peculiar properties. The first is its compulsive character: it is impossible to avoid playing with someone who is insistent. A person who has refused a challenge to play Chicken has effectively played and lost. The second peculiarity concerns the effects of commitment. A player who succeeds in making his intention to choose the risky *D* option seem convincing is bound to win *at the expense of the other player*, provided the latter is rational. This provides a game theory interpretation of the motto of the dreaded British Special Air Service (SAS): "Who dares wins". A person who enjoys a reputation of recklessness is at a decided advantage in a game of Chicken on account of the fear that he induces in any rational opponent. Long-term prisoners often cultivate such reputations for this very reason, as the following quotation from a famous ex-convict illustrates:

A con like "Harry", who had a record of violent assaults on both warders and other prisoners, could come into the T.V. room and virtually dictate what programme would be watched. With no more than a perfunctory remark such as, "What, we having the film on?" as he switched to the programme of his choice, he could control what thirty other cons were going to watch. On one occasion when I observed him do this, I'd just witnessed a show of hands which had voted for the programme which Harry had turned off. Nobody, however, protested at Harry's action. The next day I casually questioned some of the cons who'd acquiesced to Harry's demands. A few gave an account of what happened which accorded with the reality of the incident. "Well, Harry gets his own way 'cause he can have a right row," one replied. Other cons, though, rationalized explanations for what happened in order to protect their self-image from the implication that they'd backed down. . . . Another commented, "I couldn't give a fuck what was on. Like the film or the football, what's the difference?". Yet this con, like the others, had voted for the football which Harry had decided wouldn't be watched (McVicar, 1981, pp. 225–6).

If a game of Chicken is repeated a number of times, a player who gains an early advantage is likely to maintain and increase it: nothing succeeds like success in the field of brinkmanship. A player who has successfully exploited another gains confidence in his ability to get away with the risky strategy in the future, and makes his opponent all the more fearful of deviating from the cautious minimax option.

A third peculiarity of Chicken revolves around what Daniel Ellsberg (cited in Schelling, 1960, p. 13) has called "the political uses of madness". If a player is seen by his opponent to be "irrational", "not in control of himself", or frankly "mad", he gains a paradoxical advantage in a game of Chicken: people tend to give a wide berth to a madman. The following imaginary example, in which a player of "automobile Chicken" deliberately reduces his own level of rationality in order to create fear in his opponent and thereby gain a strategic advantage, is taken from Herman Kahn's book *On Escalation*:

> The 'skillful' player may get into the car quite drunk, throwing whiskey bottles out of the window to make it clear to everybody just how drunk he is. He wears very dark glasses so that it is obvious that he cannot see much, if anything. As soon as the car reaches high speed, he takes the steering wheel and throws it out of the window. If his opponent is watching, he has won. If his opponent is not watching, he has a problem; likewise if both players try this strategy (Kahn, 1965, p. 11).

Small children are remarkably adept at the political uses of madness, and so are some inmates of psychiatric hospitals. Even at the level of international politics, Adolf Hitler won a series of important games of Chicken largely on account of his apparent irrationality in the eyes of Western European leaders. It is ironical that game theory, a discipline devoted to the logic of rational decision making, should provide a logical justification for irrationality.

The game of Chicken has been used to model a variety of incidents in national and international politics involving bilateral threats. One encounter between the United States and the Soviet Union with the strategic structure of Chicken, the Cuban missile crisis of October 1962, brought the world to the brink of a nuclear holocaust. The crisis was precipitated by the discovery of Soviet nuclear missile emplacements in Cuba. The two major options considered by the United States were (a) to mount a naval blockade of the island, and (b) to launch a "surgical" air strike to knock out the missiles. The

Soviet leaders had to choose between (a) withdrawing, and (b) maintaining their missiles. A model of this situation, based on game theory analyses by Howard (1971, pp. 181–4) and Brams (1975, pp. 39–47; 1976, pp. 114–26) is shown in Matrix 6.5.

Matrix 6.5

| | | USSR | |
		Withdraw	Maintain
US	Blockade	3,3 Compromise	2,4 Soviet victory
	Air strike	4,2 US victory	1,1 Nuclear war

In the event, the United States chose the naval blockade and the Soviet Union chose to withdraw its missiles. On 28 October the crisis ended and the (immediate) danger of a thermonuclear war was averted. Dean Rusk, the United States Secretary of State, commented revealingly: "We're eyeball to eyeball, and the other fellow just blinked" (quoted in Cohen and Cohen, 1980, p. 289).

It is obvious that the same strategic structure underlies many industrial disputes in which a failure on the part of either management or workers to yield leads to an outcome, such as the bankrupting of the firm, that is disastrous for both sides. There is no completely convincing rational solution to such disputes, or to other Chicken-type strategic interactions. An interesting approach to the problem is, however, outlined in Section 6.8.

6.6   Prisoner's Dilemma

The last and most interesting of the four archetypal 2 × 2 games, and the one which has generated by far the most empirical research activity, is shown in Matrix 6.6.

Matrix 6.6
Prisoner's Dilemma

| | | II | |
		C	D
I	C	3,3	1,4
	D	4,1	2,2

Attention was first drawn to this peculiar game by Merrill Flood in 1951 and it was later explicitly formulated and given the name "Prisoner's Dilemma" by Albert W. Tucker. The name derives from the following imaginary strategic interaction to which it corresponds. Two people are arrested and charged with involvement in a serious crime. They are held in separate cells and prevented from communicating with each other. The police have insufficient evidence to obtain a conviction unless at least one of the prisoners discloses certain incriminating information. Each prisoner is faced with a choice between concealing information from the police (C) and disclosing it (D). If both conceal, both will be acquitted (3,3). If both disclose, both will be convicted (2,2). If only one prisoner discloses the information, he will not only be acquitted but will receive a reward for giving Queen's evidence while the "martyr" who conceals the information will receive an especially heavy sentence from the court (4,1 or 1,4). These payoffs are assumed to take into account the players' moral attitudes towards obstructing the course of justice, betraying a comrade, and so on; for some people the payoffs would not correspond to the Prisoner's Dilemma game. It is customary to interpret the C strategies as cooperative and the D strategies as defecting choices.

The Prisoner's Dilemma game presents a genuine paradox. The minimax strategies intersect in the D, D outcome which is the only equilibrium point in the game: neither player has any cause to regret a minimax choice if the other also chooses minimax. The minimax strategies are also dominant for both players since each receives a larger payoff by choosing D than by choosing C against either counter-strategy on the part of the other player. In other words, it is in the interest of each prisoner to disclose the incriminating evidence *whatever* the other prisoner does. But—and this is the paradox—if both players adopt this individualistic approach, the outcome (2,2) is worse for both of them than if they chose their inadmissible (dominated) strategies (3,3). In game theory terminology, the dominant strategies—which are also minimax strategies—intersect in a deficient equilibrium point.

In the logic of this game there is a curious clash between individual and collective rationality. According to purely individualistic criteria, it is clearly rational for both players to choose their D strategies, but if both opt to be "martyrs" by choosing C, the outcome is preferable for both. What is clearly required in order to ensure a better outcome for both is some principle of choice based on collective interests.

The best-known principle of this type is the Golden Rule of Confucius which found its way into the New Testament: "Whatsoever ye would that men should do to you, do ye even so to them" (Matt. vii:12). But the naïvety of the Golden Rule is evident; one of George Bernard Shaw's "maxims for revolutionists" is "Do not do unto others as you would that they should do unto you. Their tastes may not be the same." Shaw's point is illustrated by the Battle of the Sexes (Matrix 6.3), for example, in which the players have different tastes regarding evening entertainments. If both players adopt the Golden Rule by playing the "hero", the outcome is the worst possible one for both of them: each spends a lonely evening out attending a disliked entertainment.

A more sophisticated principle of collective rationality is embodied in the categorical imperative of Immanuel Kant (1785): "Act only on such a maxim through which you can at the same time will that it should become a universal law" (p. 52). The categorical imperative offers no solution to games with asymmetric equilibrium points in which the players need to choose differently from each other. Applied to the Prisoner's Dilemma game, however, it ensures that both players will choose cooperatively. But the fact cannot be escaped that this cooperative "solution" is not an equilibrium point and is therefore unstable in the sense that both players are tempted to defect from it. From an individual point of view, there is simply no rational justification for being a "martyr" through making a cooperative choice. A defecting choice not only ensures a better payoff irrespective of what the other player does, but also has the minimax property of ruling out the possibility of the worst payoff, in other words it maximizes a player's security level.

The Prisoner's Dilemma game raises subtle problems of trust and suspicion. A player who trusts the other to behave cooperatively has a reasonable justification for doing likewise, but one who suspects that the other may defect has only one reasonable course of action, namely to defect. For this reason there is a tendency for a pair of players to adopt increasingly similar patterns of play when the game is repeated a number of times. Exactly the opposite tendency is built into the dynamic structure of the game of Chicken, as explained earlier, on account of its asymmetric equilibrium points. The payoff matrices of the two games look remarkably similar to the untutored eye, but they could hardly be more dissimilar in their strategic properties. Psychologists who use these matrices in experimental research unfortunately sometimes betray a superficial grasp of such theoretical matters.

Although it is customary to interpret a C choice in the Prisoner's Dilemma game as cooperative and a D choice as competitive, it is important to bear in mind that cooperation necessarily involves an element of risk. Players are likely to be cooperative, if they are rational, only in an atmosphere of mutual trust. Defecting choices are relatively safe and may be motivated by suspicious defensiveness rather than competitiveness. Whether or not it is sensible to trust the other player in a strategic interaction depends ultimately on the circumstances in which people typically behave in trustworthy or untrustworthy ways. This is a purely empirical question which needs to be investigated by experimental methods. The relevant experimental findings are outlined in Chapter 7.

Everyday strategic interactions often turn out, upon analysis, to correspond to the Prisoner's Dilemma game. Lumsden (1973) showed empirically that the preferences of Greek and Turkish Cypriots make the Cyprus conflict a Prisoner's Dilemma game. The following example was discovered by Rapoport (1962) in Puccini's opera *Tosca*. Tosca's lover has been condemned to death, but the police chief, Scarpia, offers to save his life by instructing the firing squad to use blank cartridges in return for Tosca's sexual favours. Each protagonist faces a choice between keeping his or her end of the bargain or double-crossing the other. The payoff structure of this situation evidently corresponds to Matrix 6.6. Predictably, Tosca and Scarpia both opt for their (dominant) double-crossing strategies: she stabs him as he is about to embrace her and he turns out not to have given the order to the firing squad to use blank cartridges. The tragic outcome could have been avoided if each had trusted the other.

The Prisoner's Dilemma game is a standard model of the arms race between the superpowers (see e.g. Brams, 1975, pp. 30–9; 1976, pp. 81–91; Hamburger, 1979, pp. 76–9, 96–9). The United States and the Soviet Union each face a choice between limiting or increasing nuclear arms production. A simplified model of this situation is shown in Matrix 6.7. This example illustrates in the most vivid way imaginable the total impotence of the minimax and dominance principles as prescriptions for rational choice in mixed-motive games. If the superpowers adhere to these principles, then the world is set on a course which leads towards megadeath. To describe this outcome as "rational" is to do violence to the ordinary meaning of this word. No rational solution based on selfish philosophies is possible, but the theory of metagames discussed in Section 6.8 offers a possible—though somewhat debatable—escape from the terrible paradox of the Prisoner's Dilemma.

Matrix 6.7
The nuclear arms race

| | | USSR | |
		Limit production	Increase production
US	Limit production	3,3 Status Quo	1,4 Advantage to USSR
	Increase production	4,1 Advantage to US	2,2 Possible nuclear holocaust

6.7   Comparison of the Archetypal 2 × 2 Games

A generalized paradigm of symmetric 2 × 2 games is illustrated in Matrix 6.8. The four symmetric games that lack optimal equilibrium points can be defined more compactly by specifying the ordinal inequalities which are satisfied by each:

Leader: $T > S > R > P$;
Battle of the Sexes: $S > T > R > P$;
Chicken: $T > R > S > P$;
Prisoner's Dilemma: $T > R > P > S$.

In this notation, T represents the *temptation* to defect from the C,C outcome, R is the *reward* for joint C choices, P is the *punishment* for joint D choices, and S is the *sucker's* payoff (or in the case of Battle of the Sexes the *selfish* player's payoff) for choosing C while the other defects.

This method of defining the games has the advantage of making clear the fact that the criteria of classification depend on the ordinal properties of the payoff structures. When the payoffs are measured on interval scales, an infinite number of instances can be found of each archetypal game. If, for example, $T = 10$, $R = 5$, $P = -5$, and $S = -10$, the game is a Prisoner's Dilemma, since $10 > 5 > -5 > -10$.

A great deal of confusion about these archetypal structures exists in the experimental gaming literature. Not only are games frequently misidentified, but important strategic differences tend to be glossed over. It is not uncommon, for example, to find D choices being described (and interpreted) as competitive in a game whose underlying structure corresponds to the Battle of the Sexes. This is, of course, nonsensical because a D choice in this game can never benefit a player more than it benefits his opponent.

There are, to be sure, certain strategic resemblances between the four archetypes. Leader, Battle of the Sexes, and Chicken, in particular, share a number of features in common. In each of these three games there is a "natural" outcome which results if both players cautiously opt for their minimax (C) strategies, and in each case this outcome is a non-equilibrium point which is vulnerable in the sense that both players are tempted to deviate from it. Secondly, each of the three games possesses two asymmetric equilibrium points neither of which is strongly stable because the players disagree about which is preferable. Thirdly, none of the three games possesses a dominant strategy for either player. Finally, in all three games the worst outcome for both players results if both choose their non-minimax (D) strategies. The Prisoner's Dilemma game possesses none of the above characteristics.

There are, in addition a number of significant strategic differences between Leader, Battle of the Sexes, and Chicken. In Leader, a player who

Matrix 6.8

		II	
		C	D
I	C	R,R	S,T
	D	T,S	P,P

deviates unilaterally from minimax rewards both himself and the other player, but himself more than the other. It is for this reason that Rapoport (1967b) described a unilateral defector in this game as a "leader". In Battle of the Sexes, a unilateral defector rewards the other player more than himself, and is therefore described as a "hero". In Chicken, a unilateral defector rewards himself and punishes the other player, and is therefore called an "exploiter". Joint defection results in the worst possible outcome for both players in all three games, and that is what makes them problematical. As Rapoport pointed out, there is no room for two leaders, two heroes, or two exploiters. The reason, presumably, is that a leader needs a follower, a hero needs an admirer, and an exploiter needs a victim.

The Prisoner's Dilemma is quite different from the other three. In this game the minimax (D) strategies intersect in the only equilibrium point (P,P) from which neither player is motivated to deviate unilaterally although both benefit from joint deviation. In contrast to the other three games, both players possess dominant strategies which reinforce the individual rationality of choosing the minimax option. A player who deviates unilaterally from minimax *punishes* himself and *rewards* the other—a complete inversion of the Chicken case. Such a player may therefore be described as a "martyr". But if both players choose their non-minimax, dominated strategies, the outcome (R,R) is better for both. There *is* therefore room for two martyrs!

In all four games a minimax choice may be motivated by cautiousness and a non-minimax choice necessarily entails an element of risk. The cautious strategies in Leader, Battle of the Sexes, and Chicken are C choices, while in Prisoner's Dilemma caution dictates a D choice. Competitiveness, defined as an attempt to gain at the expense of the other player, leads to D choices in Leader, Chicken, and Prisoner's Dilemma, and to C choices in Battle of the Sexes, and cooperativeness reverses these tendencies. Cautious and competitive motives therefore conflict in Leader and Chicken, while cautious and cooperative motives conflict in Battle of the Sexes and Prisoner's Dilemma.

In each game it is in a player's interest to instil a certain attitude in his opponent. In Leader and Chicken, a player benefits by instilling *fear* in his opponent by—for example—appearing irrevocably committed to a *D* choice. In Battle of the Sexes, a player is at an advantage if his opponent experiences a sense of *resignation* through believing that he is committed to a *C* choice. In Prisoner's Dilemma it is helpful to a player if his opponent acquires a feeling of *trust*, believing that he will risk a *C* choice, and both players benefit when the feeling of trust is mutual.

6.8 Metagame Theory

One of the most important recent developments in game theory is the theory of metagames originated by Nigel Howard (1971, 1974). Many game theorists believe that metagame theory provides a rational solution to the Prisoner's Dilemma and to other mixed-motive games, but some authorities remain unconvinced. It is, perhaps, too early to reach a final verdict on metagame theory, but the fundamental ideas can be presented quite simply for the reader to judge.

An embryonic version of the metagame approach can be found in von Neumann and Morgenstern's (1944) classic exposition of game theory (pp. 100–5). The basic idea involves the construction of a model which transcends the strategies of a basic game. In the metagame model, each player is assumed to choose from among a set of *metastrategies* which are conditional upon the expected choices of the other player. These metastrategies could be given to a referee who, after examining them, would be able to make the necessary choices on behalf of the players in accordance with their wishes.

The first-level metagame of the Prisoner's Dilemma (Matrix 6.6) is constructed as follows. Player I is assumed to have the same pure strategies as in the basic game, that is, *C* and *D*. Player II, however, is assumed to choose from among the following four metastrategies:

(1) choose *C* regardless of what I is expected to choose;
(2) choose *D* regardless of what I is expected to choose;
(3) choose the same basic strategy as I is expected to choose;
(4) choose the opposite basic strategy to what I is expected to choose.

The structure of this preliminary metagame is shown in Matrix 6.9.

Matrix 6.9
First-level metagame of Prisoner's Dilemma

II

	C Regardless	D Regardless	Same	Opposite
C	3,3	1,4	3,3	1,4
D	4,1	2,2	2,2	4,1

(I on left, labelling rows C and D)

In this game there is a single equilibrium point at the intersection of row D and the column labelled "D Regardless". It corresponds, however, to the same deficient equilibrium point (2,2) of the basic game. In order to eliminate the paradox a more elaborate metagame must be constructed.

Suppose now that Player II has the same four metastrategies as in the first-level metagame, but that Player I makes a choice conditional on the metastrategies that II may choose. In this second-level metagame, Player I has 16 metastrategies based on II's possible choices. One of these metastrategies, for example, is "Choose C regardless of which of the four metastrategies II is expected to choose"; this metastrategy is denoted by $C/C/C/C$. Another is "Choose C unless II is expected to choose 'Opposite', in which case choose D"; this is denoted by $C/C/C/D$. The complete second-level metagame of Prisoner's Dilemma is displayed in Matrix 6.10.

This metagame turns out, like the first, to have an equilibrium point corresponding to the one in the basic game. It is located at the intersection of row "$D/D/D/D$" and column "D Regardless". This outcome (2,2) shows what happens if Player I opts for the metastrategy "Choose D regardless of what metastrategy II is expected to choose" and II opts for "Choose D regardless of what I is expected to choose"; obviously both players in this case are bound to choose D in the basic game.

The most important features of Matrix 6.10, however, are two new equilibrium points that have emerged where rows $C/D/C/D$ and $D/D/C/D$ intersect with the column labelled "Same". These two outcomes correspond to joint cooperative (C, C) choices in the basic game. The first is one in which Player I says, in effect, "I'll choose C if I expect you to choose either C regardless or the same basic strategy as you expect me to choose, otherwise I'll choose D", and Player II says "I'll choose the same basic strategy as I expect you to choose". In the equilibrium associated with row $D/D/C/D$, Player I says "I'll choose C if and only if I expect you to choose the same basic strategy as you expect me to choose" and Player II says "I'll choose the same basic strategy as I expect you to choose".

The equilibrium points represent rational outcomes in the sense that neither player would regret his choice of metastrategy when the other's became apparent. If Player I chooses $C/D/C/D$, for example, II can do no better than choose "Same" and vice versa. Given that Player I has chosen $C/D/C/D$, II's payoff is 3 if he chooses "Same" and he cannot improve on this; he may receive a less desirable payoff if he chooses a different column such as "D Regardless" (which yields him a payoff of only 2). Analogously, Player I cannot do better than choose $C/D/C/D$ if II chooses "Same", and he may receive a payoff of only 2 instead of 3 if he chooses, for example, $D/D/D/D$. The equilibrium points represent rational metastrategies, according to metagame theory, because neither player can improve upon his payoff—and may do worse—by deviating from such a point while the other player sticks to it.

Matrix 6.10
Second-level metagame of Prisoner's Dilemma

	C Regardless	D Regardless	Same	Opposite
C/C/C/C	3,3	1,4	3,3	1,4
C/C/C/D	3,3	1,4	3,3	4,1
C/C/D/C	3,3	1,4	2,2	1,4
C/D/C/C	3,3	2,2	3,3	1,4
D/C/C/C	4,1	1,4	3,3	1,4
C/C/D/D	3,3	1,4	2,2	4,1
C/D/C/D	3,3	2,2	3,3^e	4,1
D/C/C/D	4,1	1,4	3,3	4,1
C/D/D/C	3,3	2,2	2,2	1,4
D/C/D/C	4,1	1,4	2,2	1,4
D/D/C/C	4,1	2,2	3,3	1,4
C/D/D/D	3,3	2,2	2,2	4,1
D/C/D/D	4,1	1,4	2,2	4,1
D/D/C/D	4,1	2,2	3,3^e	4,1
D/D/D/C	4,1	2,2	2,2	1,4
D/D/D/D	4,1	2,2^e	2,2	4,1

^{e}equilibrium points

The problem remains of choosing between the three equilibrium points in the second-level metagame. The deficient equilibrium in the bottom row of the matrix can be eliminated from consideration because it yields a worse payoff to each player than the other candidates. If each player assumes that the other will pursue his own rational self-interests, then Player II will choose the column labelled "Same" and Player I will be free to choose between rows C/D/C/D and D/D/C/D. Player I's choice turns out to be straightforward because the second of these rows dominates the first. That is to say metastrategy D/D/C/D yields a payoff at least as good as metastrategy C/D/C/D whatever metastrategy II may choose, and a better payoff in one contingency (if II chooses "C Regardless").

The rational choices are thus unambiguously prescribed by metagame theory. Player I should choose $D/D/C/D$ and Player II should choose "Same". These choices result in both players making cooperative C choices in the basic game. The outcome is a payoff of 3 to each player, which is the maximum possible joint payoff in the game. The paradox of the Prisoner's Dilemma appears to have been dissolved and individual and collective rationality to have been reconciled.

The metagame could have been constructed the other way around, by assigning four metastrategies to Player I and 16 (meta)metastrategies to Player II. Howard (1971) has shown that the end result is the same in both cases. He has also managed to prove that the solution remains the same if the matrix is expanded further to take account of Player II's expectations regarding Player I's 16 metastrategies. There is an infinite hierarchy of metagames, each one larger than its predecessor, starting from any basic game; but a two-person game, if it is expanded beyond two levels, does not give birth to any new equilibrium points. In general, all of the meta-equilibria in an n-person game emerge in the nth metagame.

Whether or not metagame theory provides a solution to the Prisoner's Dilemma and to the logical paradox embodied in the game is a matter of heated debate. Howard believes so, and so does Anatol Rapoport (e.g. 1967a) who has described the approach as an "escape from paradox". The first to challenge this view was Richard Harris (1968), and a lengthy debate ensued in the pages of *Psychological Reports* throughout 1969 and 1970 with comments, replies, notes, and rebuttals from the "paradox lost" and "paradox regained" factions. A different challenge to metagame theory was later mounted by Robinson (1975) and the debate shifted to the journal *Behavioral Science*. My own view is that the theory is of enormous potential power and its most important fields of application are probably yet to be discovered, but it does not appear to provide a solution in the usual sense to the Prisoner's Dilemma game. The metagame solution depends upon each player being able to predict the other's choice, and in practice a player may not know what to expect from the other. Strictly defined, a metagame is "the game that would exist if one of the players chose his strategy after the others, in knowledge of their choices" (Howard, 1971, p. 23) and a metagame solution is the outcome which results "from each player choosing his strategy on the assumption that the other players will choose the strategies they do choose" (*op. cit.*, pp. 50–1). But metagame theory does not help a player to choose a strategy in the basic game if he does not know what the other will do. The arguments on both sides of the debate are extremely subtle and it is beyond the scope of this book to do justice to them. What is beyond question, however, is that the metagame approach throws new light on the logic of mixed-motive interactions.

It is worth making a few comments about the application of metagame theory to games other than Prisoner's Dilemma. In the game of Chicken (Matrix 6.4) the method of analysis described above reveals a metagame equilibrium point corresponding to joint C choices in addition to the unstable equilibria of the basic game (Howard, 1971, pp. 181–4; Brams, 1976, pp. 114–26). The "compromise" equilibrium corresponds to Player I's choice of $D/D/C/D$ and II's choice of "Same", as in Prisoner's Dilemma. If the Cuban missile crisis is modelled by a game of Chicken, the following unexpected conclusions emerge from the metagame analysis (Howard, 1971, p. 184):

(a) for the compromise (3,3) outcome to be stable, both of the superpowers must be willing to risk nuclear war;
(b) if one superpower but not the other is willing to risk nuclear war, that superpower wins;
(c) if neither superpower is willing to risk nuclear war, no stable outcome is possible.

These conclusions are surprising in view of the fact that the C strategies ("Blockade" and "Withdraw") in the basic game (Matrix 6.5) appear to produce a safe outcome. This safety is illusory, however, because the C,C outcome is not an equilibrium point in the basic game. The "compromise" metagame equilibrium—which leads to C choices in the basic game—is stable only because if one player refuses to give in, the other will precipitate nuclear war rather than give in. The players choose C only if they expect each other to opt for the prescribed metastrategies. The metagame analysis reveals an appalling aspect of the nuclear "balance of terror".

   In practice, the players' behaviour in mixed-motive interactions is often strongly affected by what they expect their opponents to do. It follows from this that a player can sometimes gain an advantage by misrepresenting his intentions. Metagame theory provides an extremely efficient method for investigating the logic of such strategic deception and this may turn out to be an especially fruitful area of theoretical development. An example of this type of analysis has been given by Richelson (1979), who showed that the United States and the Soviet Union both have incentives to engage in strategic deception. The Soviet Union, by pretending that it is committed to an *all-out* nuclear attack in response to a United States policy of *limited* nuclear operations, can hope to force the United States to respond with *conventional* weapons to a Soviet invasion of Europe. If, on the other hand, the United States manages to convince the Soviet Union that it is committed to the use of limited nuclear operations in response to a Soviet invasion of Europe, then the Soviet Union would have an incentive to respond with limited nuclear operations rather than an all-out nuclear attack.

   Osgood (1962) has proposed "an alternative to war or surrender" based on a *graduated reciprocation in tension-reduction* (GRIT). A nation adopting the GRIT strategy would unilaterally promise a series of low-risk actions, calculated to benefit its adversary. These promises would be unconditional, but would be accompanied by requests for some reciprocal moves on the part of the other nation. The purpose of GRIT is to foster trust between initially hostile and suspicious antagonists with the ultimate aim of avoiding war. There is some evidence that it is effective in reducing tension in two-person laboratory conflicts (for example, Tedeschi, Bonoma, and Lindskold, 1970), but whether it would work at an international level, as Osgood believes, is an open question. It would involve a degree of risk, but it could hardly be riskier in the long run than the graduated reciprocation in tension-*increase* that has taken place over the past few decades.

6.9   Summary

The chapter opened with a discussion of the mixed-motive character of variable-sum games which, in contrast to zero-sum games, generate intrapersonal as well as interpersonal conflicts. Section 6.2 outlined a classification, based on ordinal preference rankings, of all symmetric, two-person, two-choice games. Formal and informal properties of the four archetypal 2 × 2 games—Leader, Battle of the Sexes, Chicken, and Prisoner's Dilemma—were discussed in Sections 6.3, 6.4, 6.5, and 6.6 respectively. Several real-world examples, from manoeuvring in traffic to the Cuban missile crisis and the arms race between the superpowers, were used to illustrate the fundamental ideas. In Section 6.7 some of the major similarities and differences between the four archetypal games were commented upon. Section 6.8 was devoted to an outline of recent metagame theory, the possible solution that it provides to the Prisoner's Dilemma game, and the surprising conclusions about the game of Chicken to which it leads.

7

Experiments With Prisoner's Dilemma and Related Games

7.1 The Experimental Gaming Literature

FROM the early 1950s until the mid-1970s, experimental gaming research centred chiefly on two-person, mixed-motive games, especially the Prisoner's Dilemma game. It is not difficult to account for the particular attraction of researchers to these games. Apart from the factors underlying the general appeal of experimental games (see Section 5.1), they provide simple methods for investigating interesting and important aspects of strategic interaction, such as cooperation and competition, trust and suspicion, and individualism and collectivism, which are difficult or impossible to study by other means.

A necessary assumption, of course, is that behaviour in complex strategic interactions in everyday life depends, to some significant degree at least, on the principles governing behaviour in idealized and artificial laboratory games. The same kind of assumption underlies the idealized gases studied by physicists and the deliberately artificial experimental conditions created by engineers, but its cogency in the study of social interaction has not gone unchallenged. From the mid-1970s onwards, the ecological validity of simple experimental games was increasingly often called into question, and the balance of research activity began to shift towards multi-person social dilemmas which appear to model a wider range of naturally occurring strategic interactions.

Well over 1000 experiments based on the Prisoner's Dilemma and related games have appeared in print, and it would be futile to try to summarize them comprehensively in this chapter. A number of fairly comprehensive surveys have been published, and a reader who wishes to gain a thorough grasp of the literature may find the following "review of reviews" helpful.

The first published review of experimental games, which incorporated eight early experiments with the Prisoner's Dilemma and related games, was that of Rapoport and Orwant (1962). Gallo and McClintock's (1965) review was the first to concentrate specifically on mixed-motive experimental games, of which approximately 30 had been published by the mid-1960s. Both of these early reviews provide clear and informative summaries of the early findings. A number of other review articles appeared during the 1960s, and a short while later the first books devoted exclusively to mixed-motive experimental games were published; they were by Swingle (1970b), Wrightsman, O'Connor, and Baker (1972), and Tedeschi, Schlenker, and Bonoma (1973). The second of these books covers the widest area, and contains a comprehensive bibliography of approximately 1100 items on "cooperation, competition, and exploitation in mixed-motive games and coalition-formation situations".

Sixteen of the items in the Wrightsman, O'Connor, and Baker bibliography are reviews of the literature in the conventional sense (as opposed to the liberal sense in which compilers classified the entries). A large number of reviews, both general and specialized, have appeared since then, and most have adopted a more critical tone than the earlier ones. Among the most useful are those by Nemeth (1972), Apfelbaum (1974), Davis, Laughlin, and Komorita (1976), Pruitt and Kimmel (1977), Schlenker and Bonoma (1978), and Eiser (1980, chap. 7). Several specialized reviews are referred to in the following pages.

This chapter contains an introductory review of some of the more interesting and important findings on the Prisoner's Dilemma and related games. The findings are discussed under the following broad headings: strategic structure (Section 7.2); payoffs and incentives (Section 7.3); circumstances of play (Section 7.4); responses to programmed strategies (Section 7.5); sex differences (Section 7.6); attribution effects (Section 7.7); and investigations of ecological validity (Section 7.8). In Sections 7.9 and 7.10 two new experiments which bear on the question of ecological validity are reported, and Section 7.11 is devoted to a brief summary of the chapter.

7.2 Strategic Structure

Two-person, mixed-motive experimental gaming has focused mainly on the simplest symmetric 2×2 games. Apart from the Prisoner's Dilemma game, most attention has been paid to Chicken and to a theoretically trivial Maximizing Difference game.

The experimental evidence concerning behaviour in Leader, Battle of the Sexes, and asymmetric 2×2 games is scanty but interesting. Guyer and Rapoport (1972) investigated one-off choices in a variety of these neglected games and found that the majority of subjects tended to opt for their minimax strategies. The percentage of minimax choices ranged from 74 to 92 in these sundry games. Differences across games were largely explicable on the basis of informal game theory analyses. Fewer subjects deviated from minimax, for example, in games in which minimax strategies were also dominant, which is entirely consistent with common sense.

Matrix 7.1
Generalized 2 × 2 payoff matrix

II

		C	D
	C	R,R	S,T
I			
	D	T,S	P,P

In Leader, defined according to the inequalities $T > S > R > P$ (see Matrix 7.1), 78 per cent of choices were minimax (C), i.e. the majority of subjects cautiously avoided the worst possible outcome which might arise from a D choice. In Battle of the Sexes (which the authors call "Hero"), defined by $S > T > R > P$, the percentage of minimax (C) choices was 82; in other words more than four out of five subjects followed the minimax principle rather than the Golden Rule. One of the games included in this investigation was a version of the Maximizing Difference game ($R > T > S > P$) in which neither player can benefit by deviating from minimax (C) *whatever* the other player does. In this game 90 per cent of subjects chose the minimax strategy, and the investigators attributed the remaining 10 per cent to simple errors on the part of the subjects. As we shall see, subjects' choices are less predictable when games are played more than once.

Turning now to the Prisoner's Dilemma game (PDG), the most striking general finding is undoubtedly the *DD lock-in effect*. In addition to the defining inequalities $T > R > P > S$, most experimental PDG experiments have used matrices in which $2R > S + T$ so that coordinated alternations between CD and DC outcomes are less profitable to the players than CC outcomes. When the game is repeated many times there is a tendency for long runs of D choices by both players to occur. Although the D strategy is dominant and corresponds to minimax in this game, both players are better off if both choose C. Luce and Raiffa (1957) therefore predicted that repetitions of the PDG would lead to CC joint cooperative outcomes: "We feel that in most cases an unarticulated collusion between the players will develop. . . . This arises from the knowledge that the situation will be repeated and that reprisals are possible" (p. 101). A year after the publication of Luce and Raiffa's book, however, Flood (1958) reported that the DD lock-in is a common occurrence. This finding was replicated by Scodel *et al.* (1959) and has since then been confirmed literally hundreds of times. Mathematical game theorists, however, remain largely ignorant of these experimental findings.

The following remark from a recent mathematical text is not untypical: "If . . . the game is played *many* times then it is arguable that collective rationality is more likely to come into play. For although the prisoners cannot communicate directly, they can signal their willingness to cooperate with each other by playing cooperative strategies" (Jones, 1980, p. 78).

Some illumination of the DD lock-in effect has been provided by studies in which the game has been repeated literally hundreds of times. The most detailed and thorough study of this kind is undoubtedly the work of Rapoport and Chammah (1965); these researchers were the first to map the long-term time courses of choices in the PDG. Their results, which have been confirmed by several subsequent investigators, showed that three phases typically occur in a long series of repetitions. On the first trial, the proportion of cooperative (C) choices is typically slightly greater than 1/2, but this is followed by a rapid decline in cooperation (a "sobering period"). After approximately 30 repetitions, cooperative choices begin to increase slowly in frequency (a "recovery period"), usually reaching a proportion in excess of 60 per cent by trial 300.

The moderately high proportion of initial C choices has been interpreted variously as indicating an initial reservoir of goodwill or simply a naïve lack of comprehension on the part of the subjects of the strategic structure of the game. The sobering period may consequently reflect a decline in trust and trustworthiness, an increase in competitiveness, or merely a dawning understanding of the payoff matrix. The recovery period can be interpreted relatively unambiguously: it probably reflects the slow and imperfect growth of an "unarticulated collusion" between the players such as was predicted by Luce and Raiffa (1957).

Other interpretations of the motivational sources of C and D choices in the PDG are of course possible. McClintock (e.g. 1972) has distinguished between motives on the part of the players to *maximize joint payoffs* (leading obviously to C choices), to *maximize individual payoffs* (for which no clear strategy is available), and to *maximize relative payoffs*, that is, to "beat the opponent" (leading to D choices). McClintock has pointed out that, because of the motivational ambiguity regarding the appropriate strategy for individual payoff maximization, the interpretation of behaviour in the PDG is inherently problematical. To be specific, C choices may reflect attempts to maximize joint payoffs *or* individual payoffs (on the basis of trust), and D choices may be made with a view to maximizing individual payoffs (on the basis of mistrust) *or* relative payoffs.

When other factors are held constant, the game of Chicken (with $T > R > S > P$, and $2R > S + T$ to make coordinated alternation less profitable than joint C choices) tends to elicit higher proportions of C choices than the PDG. This is in line with predictions based on informal considerations, since in the PDG the D strategy is both dominant and relatively safe (minimax) while in

Chicken neither strategy is dominant and C is the safer alternative. The most thorough investigation of behaviour in Chicken was reported by Rapoport and Chammah (1969). They found that the frequency of C choices on early trials is on average about 65 per cent, followed by a very slight decrease over approximately the next 30 trials, and a steady increase from that point onwards, reaching more than 70 per cent after 300 repetitions.

The problem referred to earlier of distinguishing between motives of joint, individual, and relative payoff maximization apply equally to the game of Chicken as to the PDG. This problem does not arise, however, in the Maximizing Difference game (MDG) in which the only justification for a D choice—in so far as it can be justified at all—is relative payoff maximization. Since both players in the MDG receive their highest possible payoffs if both choose C, a D choice must presumably be motivated by a desire to "beat the opponent" although it entails certain loss to the D-chooser as well. The version of the MDG used in most experiments is defined as follows: $T = 5$, $R = 6$, $P = 0$, $S = 0$ (see Matrix 7.1). A number of investigations of behaviour in this game (reviewed by McClintock, 1972) have revealed surprisingly high frequencies of D choices, and they are usually interpreted as attempts at relative payoff maximization. McClintock and McNeel (1967), for example, found that pairs of Belgian students who were given the opportunity to win substantial sums of money over 100 trials in the MDG chose C only about 50 per cent of the time, and subjects who were offered smaller monetary rewards deviated from the C strategy even more frequently.

McClintock (1972) has interpreted these results as follows:

> In effect, the high reward subjects were still willing to forego considerable material gain which would have obtained if they had maximized joint gains to ensure that they would have more points than the other, or at least, not fall appreciably behind the other in score (p. 291).

One cannot take issue with this interpretation as far as it goes; the only other explanation for D choices is a failure on the part of the subjects to understand the payoff matrix, and evidence from other types of games makes this seem unlikely. But neither McClintock nor subsequent commentators have drawn the logical conclusion from this that some of the subjects were playing the game according to a subjective utility structure different from the specified payoff matrix. Their preferences among the possible outcomes of the game evidently did not correspond to the MDG payoff structure. For if, as McClintock implies, a subject prefers the DC to the CC outcomes, then—in terms of subjective utilities—R is not greater than T and the (subjective) game is not MDG. The findings of research in this area therefore demonstrate beyond reasonable doubt that subjects in experimental games do not invariably play according to the explicit payoff structures presented to them. This problem strikes at the heart of experimental gaming and is badly in need of direct investigation. The experiments reported in Sections 7.9 and 7.10 are addressed partly to it.

7.3   Payoffs and Incentives

Numerous experiments have centred on the effects of varying the relative magnitudes of the payoffs within a specified strategic structure. In the PDG or Chicken, for example—and the majority of experiments have been based on these two games—it is possible to vary the relative magnitudes of the four payoffs in Matrix 7.1 above, namely T (the temptation to be a unilateral defector), R (the reward for joint cooperation), S (the sucker's payoff for unilateral cooperation, and P (the punishment for joint defection) while preserving their ordinal ranking.

The results of payoff variation in the PDG have usually been in line with common-sense predictions. When R is increased relative to P, the proportion of C choices tends to increase, and when T is increased relative to S, the proportion of C choices decreases, as one might expect. Rapoport and Chammah (1965) found that a *cooperation index*, given by the ratio $r = (R - P)/(T - S)$ is a fairly reliable indicator of the proportions of C choices generated by different PDG matrices. The correlation between the (log transformed) cooperation index r and the proportions of C choices across matrices was found to be 0.641.

With regard to Chicken, Rapoport and Chammah (1969) investigated five matrices in which $T = 2$, $R = 1$, and $S = -2$, while P varied from -3 to -40. A strong relationship emerged, as expected, between the degree of dangerousness of the game, as indexed by the value of P, and the proportion of C choices. Subjects tended to behave more cautiously, that is, to choose C more frequently, the worse the punishment for joint defection. The authors appended the following comment:

> Penologists sometimes argue that certainty of conviction is a more powerful deterrent of crime than severity of punishment. Something of this sort may also be at the basis of the various theories of measured response, which have supplanted the short-lived doctrine of massive retaliation in the thinking of American strategists. Our results with Chicken seem to show the opposite: More severe punishment seems to be a more effective deterrent than more certain punishment (Rapoport and Chammah, 1969, p. 169).

In the PDG, Chicken, and other mixed-motive games, subjects generally respond in understandable ways to payoff manipulations. These findings, incidentally, seem to contradict the belief of some critics of experimental gaming that the behaviour of subjects "is primarily due to the essential incomprehensibility of the situation in which the subject is placed" (Nemeth, 1972, p. 213).

The problem of incentives has bedevilled experimental gaming ever since the 1960s when it was first suggested that subjects might not take their tasks seriously. Limited research funds and ethical considerations make it impossible in most experiments to assign anything more than small monetary rewards to the subjects' payoffs. In their early review, Gallo and McClintock (1965) suggested that the surprisingly low levels of cooperation generally found in PDG experiments might be attributable to the absence of substantial monetary incentives. Gallo's own doctoral dissertation (Gallo, 1966) was addressed to this question and his findings supported the view that large monetary incentives might generate higher levels of cooperation. But the evidence was only indirect because Gallo used a "trucking" game (see Section 7.4) whose payoff structure is quite different from the PDG.

Subsequent research involving high and low incentives (reviewed by Oskamp and Kleinke, 1970) has generated equivocal results. Some experiments on the PDG (e.g. Wrightsman, 1966) and Chicken (e.g. Sermat, 1967a, 1967b) found *no* incentive effects, others (e.g. Stahelski and Kelley, 1969) found significantly *more* C choices in high-incentive conditions, and still others (e.g. Gumpert, Deutsch, and Epstein, 1969) found *fewer* C choices in high-incentive conditions. The only reasonable conclusions seem to be (a) that monetary incentives do not affect behaviour in the PDG or Chicken in any pronounced or consistent manner, and (b) that incentives probably interact with other unknown variables in determining choices in these games. In the MDG with $T = 5$, $R = 6$, $P = 0$, and $S = 0$ (see Matrix 7.1), on the other hand, incentives do seem to be positively related to the frequency of C choices (e.g. McClintock and McNeel, 1967). This is hardly surprising in view of the fact that C choices are clearly the only rational strategies in the MDG (see Section 6.2): subjects are more likely to pursue their own interests rationally when the incentive for doing so is large.

7.4 Circumstances of Play

It is possible to present a game in a variety of apparently different but logically equivalent forms, and the evidence suggests that the mode of presentation has a powerful effect on the behaviour of experimental subjects. The most interesting findings in this area relate to *decomposed* or *separated* PDGs. A conventional PDG and two of its derivative decompositions are displayed in Matrices 7.2. The particular decompositions (Matrices 7.2b and c) are merely illustrative. If a PDG can be decomposed at all, and this is possible only if $T - R = P - S$, then there are always infinitely many ways to do so. (Chicken and other games without dominant strategies cannot be decomposed.)

Matrices 7.2
PDG and two derivative decompositions

	II	
	C	D
C	3,3	1,4
D	4,1	2,2

(a)

	Own Payoff	Other's Payoff
C	0	3
D	1	1

(b)

	Own Payoff	Other's Payoff
C	1	2
D	2	0

(c)

(In matrix (a), the left labels C and D belong to player I.)

A decomposed matrix such as 7.2b (or 7.2c) is interpreted as follows. A player's options are labelled as usual C and D. The choice of either option is shown in the corresponding row to yield a payoff to the player who chooses it and, simultaneously, a payoff to the other player. Each player chooses a row from the same decomposed matrix, so to determine a player's total payoff we must add his "own payoff"—resulting from the row he chooses—to the "other's payoff" awarded to him by his opponent's choice.

Let us examine the first decomposed Prisoner's Dilemma (Matrix 7.2b). If a subject is presented with this payoff schedule and told that the other member of the pair has an identical copy, then it is possible for him to work out the consequences of any joint decision. If he chooses C, for example, he can see that he will award himself no points and award the other subject three points. Thus if both choose C, each receives three points "from the other". This corresponds to the CC outcome in the conventional matrix (a). If both choose D, each receives one point "from himself" and one point "from the other", so each ends up with two points. The consequences of other combinations of choices can be deduced in the same way. The important point is that this decomposition (b) is mathematically equivalent to the conventional payoff matrix (a), and it is easily verified that the other decomposition (c) is likewise equivalent to (a). Each version conveys exactly the same information in a different form.

A decomposition of the kind shown in Matrix 7.2c invites an interpretation of the PDG in terms of generosity versus stinginess, as Pruitt and Kimmel (1977, p. 373) have pointed out. People who are next-door neighbours frequently have to choose between lending (C) and refusing to lend (D) each other eggs, milk, tools, baby-sitting services, and the like. There is a temptation to refuse on account of the inconvenience caused by lending, but both are better off if both agree to lend than if both refuse. Only at the level of collective rationality is virtue its own reward; at the individual level most of the benefit goes to the other person.

Several empirical studies (e.g. Evans and Crumbaugh, 1966; Pruitt, 1967) have shown that subjects are unusually willing to make C choices in decomposed PDGs such as Matrix 7.2b in which a player's payoffs appear to come largely "from the other". This form of display emphasizes a player's dependence on the other's willingness to cooperate, and it is presumably the heightened awareness of interdependence that facilitates the development of mutual cooperation and trust. A similar conclusion has emerged from naturalistic research in the field of group dynamics (e.g. Sherif *et al.*, 1961; Worchel, Andreoli, and Folger, 1977). Experiments have shown that, in a decomposed PDG that elicits high levels of cooperation, subjects are especially quick to reciprocate C choices and especially slow to defect to D in response to the other players' defecting choices (Pruitt, 1970; Tognoli, 1975).

Various non-numerical methods of presenting mixed-motive games to subjects have been tried by researchers. The purpose is usually to create a laboratory mixed-motive interaction that is less abstract and "meaningless" than the conventional payoff matrix or, for that matter, than a decomposed matrix. The most extensively researched of these non-numerical games is Deutsch and Krauss's (1960) *trucking game*. A useful review of experiments based on the trucking game can be found in Harvey and Smith (1977, pp. 311–23). The essential features of the game are as follows. Each of two subjects plays the role of a truck driver trying to get to a certain destination as quickly as possible. Each has a choice between two possible routes, one considerably shorter than the other. The short route, however, includes a stretch of single-lane traffic which the trucks have to traverse in opposite directions; thus if both subjects choose the shorter route simultaneously, a head-on blockage occurs and both lose valuable time until one decides to back away and switch to the longer route. In some conditions, one or both of the subjects controls a "gate" which can be used to prevent the other from passing beyond the single-lane stretch of the road.

Playing repeated trials of the trucking game without gates, Deutsch and Krauss's (1960, 1962) subjects quickly learned to cooperate by alternating on the shorter route. But when both subjects controlled gates—the *bilateral threat* condition—frequent blockages were typically recorded and, somewhat surprisingly, no improvement took place over trials. The *unilateral threat* condition, in which one member of the pair controlled a gate, produced intermediate levels of cooperation and efficiency. Deutsch interpreted these results as showing that the availability of potential threats encourages people to behave competitively, and he has argued (Deutsch, 1969) that this conclusion has vital implications for international relations, disarmament, and military deterrence. There is some evidence in support of Deutsch's interpretation (e.g. Smith and Anderson, 1975), but other views are possible. Kelley (1965) pointed out, for example, that the control of gates may increase the level of competition simply because it permits a player to maximize relative gain by blocking the other player in the single-lane stretch and taking the longer route. When neither player controls a gate, of course, neither can do anything to gain at the other's expense because blockages are always mutual.

Although the results of research with the trucking game are often compared with—or even assimilated to—the results of matrix gaming experiments, they ought really to be treated as coming from an independent line of enquiry of particular relevance to the study of threats. In spite of Wrightsman, O'Connor, and Baker's (1972, p, 191) claim to the contrary, the trucking game *is* a mixed-motive game, but its strategic structure is extremely complex in the threat conditions, and it does not correspond to the game of Chicken (as many believe) or to any other well-understood payoff matrix.

Before leaving the topic of circumstances of play, something needs to be said about the fact that subjects in traditional experimental games are forbidden to communicate verbally with each other. In naturally occurring mixed-motive interactions the participants are often (though by no means always) able to issue verbal threats, promises, commitments and the like. A number of commentators have suggested that the uncooperative behaviour observed in many mixed-motive gaming experiments "stems mainly from the subjects' inability to communicate their goals or preferences" (Nemeth, 1972, p. 213). Common sense certainly suggests that communication should facilitate mutually beneficial collaboration, and a number of researchers have examined this hypothesis experimentally.

One of the classic experiments in this area is that of Evans (1964). This study investigated PDG behaviour under three conditions: enforceable promises with stiff penalties for breaking them, unenforceable promises, and no communication. As expected, the highest level of *C* choices was found in the enforceable-promises condition and the lowest in the no-communication condition. When subjects are permitted to promise, threaten, lie, and deceive without restriction, it is more difficult to predict the effects. Deutsch (1958) was the first to discover that unrestricted communication leads to increases in the frequency of *C* choices only when the subjects have a particular motivational set. Deutsch's subjects received instructions calculated to encourage either a competitive, a cooperative, or an individualistic set. Opportunities for communication made little difference in the first two conditions; it was only when the subjects were instructed to think of themselves and their own interests alone that communication led to increased levels of *C* choices. This finding, which has been replicated a number of times, makes sense in the light of McClintock's (1972) theoretical analysis of the motivational bases of *C* and *D* choices in the PDG. The prescribed strategies for relative payoff maximization, which is equivalent to Deutsch's competitive set, and joint payoff maximization—Deutsch's cooperative set—are unambiguous, but individual payoff maximization has no clear implications for strategy choice. There is thus more scope for interpersonal influence in individualistically motivated pairs.

Wichman (1972) has reported the most sophisticated study of communication in the PDG. Individualistically motivated subjects were permitted either to see or hear their opponents, both to see and hear them, or neither to see nor hear them. The results indicated that subjects chose C most frequently in the see-and-hear condition, less frequently in the hear-only condition, less frequently still in the see-only condition, and least frequently of all in the isolated condition. It was only in the see-and-hear condition that subjects succeeded in establishing and maintaining high levels of joint C choices over the series of 70 trials. The results of this investigation show that both verbal and non-verbal communication can facilitate the development of cooperation and trust, and that the effects of communication are rather more subtle than was previously assumed.

7.5   Responses to Programmed Strategies

Oskamp (1971) has provided an excellent review of the evidence concerning the effects of the other player's strategy on behaviour in two-person, mixed-motive games. Experiments in this area involve pitting subjects against a simulated "other player" in the form of a computer or a confederate of the experimenter programmed in advance to make predetermined sequences of choices. In most of these experiments the subjects are led to believe that they are playing against real opponents.

The two simplest programmed strategies are those in which the other player chooses either C on every trial (100 per cent C) or D on every trial (0 per cent C). The effects of these extreme programmed strategies are quite different in the PDG and Chicken (e.g. Sermat, 1967b). In the PDG, 100 per cent C (unconditional cooperation) elicits much higher frequencies of C choices from subjects than 0 per cent C (unconditional defection). This is, of course, in line with expectations, because considerations of self-defence force a subject to choose D against the 0 per cent C program in order to avoid the worst possible outcomes (sucker's payoffs). Many subjects reciprocate the cooperation of the 100 per cent C programmed strategy, but what is surprising is the large proportion of subjects who seize the opportunity of exploiting the unconditionally cooperative program to their own advantage by choosing D. The tendency for many subjects to exploit "pacifist strategies" has been confirmed in games other than the PDG (e.g. Shure, Meeker, and Hansford, 1965) although the level of exploitation depends on a number of circumstantial factors (Reychler, 1979).

Exploitation of the 100 per cent C programmed strategy is also common in the game of Chicken, but in comparison with 0 per cent C the pattern of results is exactly the reverse of that in the PDG (Sermat, 1967b). In Chicken,

100 per cent C elicits significantly *fewer* C choices from subjects than 0 per cent C. This difference is perfectly intelligible, because Chicken differs from the PDG in that the only way to avoid the worst possible outcome against the 0 per cent C program is to choose C. The experimental evidence on the extreme programmed strategies harmonizes well with the theoretical analyses of the PDG and Chicken outlined in Sections 6.5 and 6.6.

A conditional programmed strategy which has attracted a great deal of research attention is the "tit-for-tat" (TFT) strategy in which the computer or confederate chooses, on each trial, the option chosen by the subject on the previous trial. This may be interpreted as a way of signalling to the subject "I'll cooperate if and only if you cooperate". Possibly because it seems such an eminently sensible way of playing the PDG, most reviewers (including Davis, Laughlin, and Komorita, 1976, p. 518; Pruitt and Kimmel, 1977, p. 380) claim that it elicits more frequent C choices from subjects than other programmed strategies including 100 per cent C, but a careful reading of the evidence does not support this interpretation unequivocally. The TFT strategy does, however, elicit more C choices than a randomly mixed strategy containing the same number of C choices (Crumbaugh and Evans, 1967), which proves that the *pattern* of C choices has an effect independent of the mere *frequency* of C choices in a programmed strategy.

A number of experiments have been devoted to the effects of programmed strategies involving changing frequencies of C choices over trials. Harford and Solomon (1967) were the first to examine *reformed sinner* and *lapsed saint* strategies. The reformed sinner initially chooses 0 per cent C, then switches to 100 per cent C, and adopts TFT in the final phase. The lapsed saint begins with 100 per cent C and then switches to TFT. In the PDG, Harford and Solomon found that the reformed sinner program elicits much higher levels of C choices than the lapsed saint program, and this effect has been replicated several times.

7.6 Sex Differences

One of the most striking and unexpected findings to emerge from early mixed-motive gaming experiments was the apparent tendency for females to exhibit much *lower* frequencies of C choices than males in both the PDG and Chicken. Considering traditional sex roles, one might expect females to behave *more* cooperatively than males. In their extensive investigation of PDG behaviour, Rapoport and Chammah (1965) found an average of 59 per cent C choices in male pairs and only 34 per cent C choices in female pairs. The average percentage of C choices in mixed pairs was about 50. The sex difference was not detectable in mixed pairs, presumably because the strategic structure of the PDG encourages players to choose the same strategies as their opponents (see Section 6.6). Rapoport and Chammah's evidence suggests that, in mixed pairs, females converged towards their male opponents' more cooperative play to a greater extent than the males converged towards the females. The marked difference in same-sex pairs, which was found in all of the seven matrices used in the investigation, was due largely to a steeper and longer-lasting initial decline in C choices over trials among the females than among the males. In a later investigation, Rapoport and Chammah (1969) found an equally marked sex difference in the game of Chicken, with females once again exhibiting fewer C choices than males. In both games, the *initial* frequencies of C choices were the same among males and females.

There are now more than 100 published experiments containing data on the sex difference in mixed-motive games, but the effect is still shrouded in mystery. Many researchers have replicated the original findings in the United States (e.g. Hottes and Kahn, 1974), Britain (Mack, 1975) and elsewhere, but some have found no significant sex differences (e.g. Kanouse and Wiest, 1967), and a (very) few have found *greater* frequencies of C choices among females than among males (e.g. Tedeschi, Bonoma, and Novinson, 1970). It has been claimed that the sex difference, at least in the PDG, is an artifact, and some evidence suggests that it evaporates when the experimenter is female (Skotko, Langmeyer, and Lundgren, 1974), but Gibbs (1982) found a clear sex difference in two of her experiments controlled by a female experimenter. Gibbs's results suggest that the sex difference may be more pronounced with a male experimenter, but experimenters of either sex were shown to differ markedly from one another in the degree of sex difference in the subjects' behaviour that they elicited. Experimenters evidently differ widely in the extent to which they create an atmosphere encouraging sex role-related behaviour. The presence of a male experimenter seems *usually* to enhance the salience of sex roles, but a female experimenter *sometimes* has the same effect.

Why do females generally behave more competitively than males in mixed-motive games when sex roles are salient? Stated in this way, the phenomenon seems extremely counter-intuitive. But it was pointed out in Sections 6.6 and 6.7 that D choices invite a variety of alternative interpretations apart from competitiveness. It is tempting to explain the generally lower frequency of C choices among females in the PDG by the fact that D is the relatively safe strategy in this game, thus females may have a tendency to behave more *cautiously* (rather than more competitively) compared with males. It is not unreasonable to assume that, in Western industrial cultures, males are brought up to behave with more boldness, daring, and risk taking than females, and this would account for their greater willingness to choose non-minimax C choices in the PDG in spite of the danger of receiving sucker's payoffs. But this theory collapses in the light of the evidence showing the same tendency in Chicken, because the cautious policy in Chicken is to choose C, and this behaviour is more characteristic of males than of females. Hottes and Kahn (1974) argued plausibly that females respond more *defensively* to the demand characteristics of a PDG experiment than males; but why do they not manifest this greater defensiveness by choosing C more frequently in Chicken experiments?

A D choice in the PDG shares one property in common with a D choice in Chicken, and this suggests a possible clue to the sex difference mystery. In both games a D choice insures a subject against the possibility of receiving the sucker's payoff or, in other words, of receiving a worse payoff than the other player. The following hypothesis is consistent with nearly all of the evidence. Perhaps some experimenters, including probably the majority of male experimenters, induce a stronger feeling of *evaluation apprehension* in female than in male subjects. If the subjects then construe the gaming experiment as a competition in which their intelligence or skill is being judged in relation to that of their opponents, then females may choose D more frequently than males in order to rule out the possibility of losing the competition. The motive to avoid appearing more foolish than the other player may in these circumstances override the explicit payoffs in the matrix; the subjective utility of the sucker's payoff may be worse in the minds of anxious subjects than in the matrix. If this hypothesis is correct, then it is misleading to suggest that "women have a greater tendency to respond suspiciously, resentfully, conservatively, and thus competitively more than do men" (Bixenstine and O'Reilly, 1966, p. 263). One might rather say that *in certain circumstances* women are more anxious than men to avoid appearing foolish in comparison with their opponents. This is nothing more than a reasonable conjecture, but it has the virtue of being empirically testable and is consistent with both the PDG and the Chicken findings.

7.7   Attribution Effects

A series of interrelated studies by Kelley and Stahelski (1970a, 1970b, 1970c) focused on the effects of people's beliefs about their opponents' intentions in the PDG. In the course of repeated plays of the game, a person can hardly avoid attributing intentions to the other player on the evidence of the latter's behaviour. There is plenty of scope for attributional errors, however: a C choice can be misinterpreted as indicating a cooperative overture when in fact it is calculated to lure the opponent into making a C choice with a view to exploitation, or vice versa; and a D choice can be misinterpreted as a competitive act when in fact it is motivated by self-protective considerations based on an expectation that the other will choose D, or vice versa. In the PDG there are numerous other plausible motives and attributions, and it is of some importance to know how accurately people are able to infer each other's intentions.

Kelley and Stahelski (1970a) invited a group of subjects to express their own cooperative or competitive intentions to the experimenter before playing 40 trials of the PDG in pairs. After each block of 10 trials they were questioned about their inferences regarding their opponents' intentions. Attributional errors were found to relate to competitiveness in an interesting way. Subjects whose own intentions were cooperative were generally able to infer their opponents' intentions—whether cooperative or competitive—beyond the level of chance after only 10 trials. Subjects whose own intentions were competitive, on the other hand, were generally unable to infer the intentions of cooperative opponents even after 40 trials.

Kelley and Stahelski put forward an intriguing theory to account for their findings. Consider the implications of *projective* attributional errors which arise from the assumption which people often make that others are similar to themselves. Cooperative people may tend to assume, in the absence of any evidence to the contrary, that others are generally cooperative, and competitive people may tend to assume that others are generally competitive. But cooperative people are likely to discover through bitter experience in everyday mixed-motive interactions that people vary in their cooperative or competitive tendencies, and the projective fallacy is therefore likely to be corrected. Competitive people, in contrast, are less likely to have their projective attributional errors refuted by experience. The reason for this is that competitive behaviour in PDG-type interactions forces other people to respond competitively to avoid their worst possible outcomes even if they would prefer to cooperate given half a chance. The Germans have a saying: "Wie man hineinruft, so schallt es heraus" (The way you shout determines the sound of the echo).

The implication of the above line of argument is this: a person who behaves competitively and predicts that others will behave similarly tends to elicit competitive behaviour from those with whom he interacts. A classic self-fulfilling prophecy is at work, and the projective attributional errors of competitive people, far from being corrected, are likely to be confirmed and reinforced by experience.

A crucial empirical assumption built into this theory is that cooperative and competitive people hold widely differing world views. Kelley and Stahelski (1970c) produced some impressive empirical evidence in support of this assumption. Subjects who favoured competitive solutions to a wide range of problems—the PDG, bargaining conflicts, and even social–political disagreements—were found to believe, in general, that others are uniformly competitive, whereas cooperative subjects tended to believe that some people are cooperative and others competitive. This relationship between cooperative or competitive intentions and attributions of intent to others is referred to by the authors as their *triangle hypothesis* because of its visual appearance when depicted as in Table 7.1. As shown in the first row of the table, very cooperative people attribute the full range of intentions to others, whereas very competitive people, represented in the bottom row, tend to attribute only competitive intentions to others. According to Kelley and Stahelski, competitive people have a cynical and authoritarian outlook on life; they believe in *and create* a social environment of competitive and untrustworthy adversaries against whom they feel impelled to defend themselves by behaving competitively. Cooperative people, on the other hand, are less pessimistic in outlook and have a more realistic appreciation of the diversity of human nature.

Table 7.1   The triangle hypothesis

Attributions of intent to others

		V. cooperative	Neutral	V. competitive
Own intentions	V. cooperative	X	X	X
	Neutral		X	X
	V. cooperative			X

Eiser and Tajfel (1972), using a non-PDG mixed-motive game, have shown that cooperative subjects are usually more eager than competitive subjects to obtain information that might reveal the motives governing each other's behaviour. This provides indirect support to the triangle hypothesis, according to which only cooperative subjects would need to know the predispositions of others in order to formulate their own strategies; competitive subjects presumably think they know that all people are motivated like themselves. Other researchers have argued against over-extending the triangle hypothesis and have shown that there are many situational factors which need to be taken into account. Miller and Holmes (1975) and Kuhlman and Wimberley (1976), for example, have corroborated the triangle hypothesis in the PDG but found it inadequate to account for behaviour in certain other games. This is in line with common sense because it is in the PDG that choices are most open to misinterpretation.

7.8   Investigations of Ecological Validity

The experimental literature on the PDG and related games has potentially important implications for a wide range of social and interpersonal problems in everyday life. But, for reasons discussed in Section 5.4, serious doubts have been raised about the ecological validity of experimental games. Wrightsman, O'Connor, and Baker (1972) expressed surprise that "apparently no studies have compared degree of cooperative behavior in a laboratory mixed-motive game with cooperation in different real-world tasks" (p. 207). A handful of experiments, one of which was listed in Wrightsman, O'Connor, and Baker's bibliography but not commented upon in the text of their book, have addressed this problem by comparing choices in abstract, matrix games with choices in relatively lifelike laboratory tasks, and these will now be critically reviewed.

Orwant and Orwant (1970) compared the choices of 165 students of journalism in 10 conventional PDG payoff matrices with their choices in 10 "interpreted" versions of these matrices. The major finding was that "the interpreted version[s] elicited significantly more cooperation than the abstract version[s]" (p. 95). But there are several peculiarities about this experiment which make it unsafe to draw any firm conclusions from the results. Most importantly, the strategic structures of the "interpreted" versions did not correspond to those of the matrix games, and this lack of correspondence may, in part at least, explain the observed differences in choice behaviour. Even the rank order of outcome preferences was not the same in the "interpreted" and matrix versions, and some of the "interpreted" versions were not even PDGs.

Young (1977) investigated behaviour in an abstract and a "structurally equivalent" lifelike version of a mixed-motive game. He hypothesized that the realism of the lifelike version would engage norms of reciprocity and social responsibility and thus elicit more cooperative choices than the abstract version, but his results revealed exactly the reverse. Unfortunately, the "abstract" version was not entirely abstract: in both treatment conditions the subjects were told that the decisions involved a situation in which a football team captain, played by the subject, comments to a team coach in one of two alternative ways about a problem of team morale, and the coach responds in one of two alternative ways. Secondly, the strategic structures of the abstract and lifelike versions were quite different from each other: the former was an ill-defined asymmetric game and the latter was a PDG. These and other flaws in the design and statistical analysis of this experiment make it hazardous to draw any conclusions from the results.

Sermat (1970) reported four experiments in which the behaviour of subjects who had previously displayed either highly cooperative or highly competitive behaviour in the PDG or Chicken was examined in a specially constructed "Paddle game" or in a picture-interpretation task. The comparison was *not*, however, made within the same experiment with irrelevant variables held constant. The results showed that, in contrast to their behaviour in the matrix games, nearly all of the subjects were highly cooperative in the Paddle game. There was very little relationship between behaviour in the matrix games and behaviour in the Paddle game, and no relationship was found between game behaviour and picture-interpretations. The Paddle game, which involved manoeuvring a stick backwards and forwards in a slot, "was assumed to be relatively similar to conventional mixed motive-games like the Game of Chicken" (p. 94). In reality, however, the strategic structures of the two games were utterly different: the Paddle game turns out, on analysis, to correspond to a sequence of 3×3 zero-sum games, while Chicken is a 2×2 mixed-motive game. Sermat expressed an awareness of the shortcomings of his investigation; but he argued that, if evidence for a relationship between laboratory game behaviour and behaviour in other situations is not forthcoming, then "the theoretical contribution of game research may have to be stated in other terms than its relevance to interpersonal behavior in real-life situations" (p. 108).

Eiser and Bhavnani (1974) investigated the behaviour of 80 subjects over 10 trials in a PDG against a programmed TFT strategy from the other player: on the first trial, the subject received a C choice from the other player, who thereafter copied on each trial the subject's previous choice. The decision context was similar in all treatment conditions: subjects made their choices on the basis of an identical payoff matrix. The instructional set was, however, varied in a way that introduced contextual variations in the subjects' *perceptions* of the decision contexts. One group were given no contextual information beyond the abstract rules of the game, another were told merely that it was a simulation of economic bargaining, another thought it had to do with international negotiations, while a fourth group were led to believe that it concerned friendly or unfriendly interactions between pairs of individuals.

The investigators hypothesized that the interpretation of the game situation in terms of economic bargaining would tend to engage competitive motives and lead to more D choices than the "international" and "interpersonal" treatment conditions, since the first "provides an excuse for exploitative self-interest" while in the other two types of situation "cooperation is more highly valued" (Eiser and Bhavnani, 1974, p. 94). The hypothesis was confirmed: the frequencies of C choices in the abstract and "economic" treatment conditions were fairly typical of frequencies reported in the literature on the PDG, but the frequencies of C choices in the "international" and "interpersonal" conditions were significantly higher. The authors concluded that "extrapolations from the results of PDG experiments to particular kinds of real-life situations must depend for their validity at least partly on whether the subjects themselves interpret the game as symbolic of the situations in question" (p. 97). Although Eiser and Bhavnani's experiment was nicely conceived and competently executed, and although their results were highly suggestive, no attempt was made to use radically different experimental tasks or to make any of the conditions genuinely lifelike.

The experiments described in Sections 7.9 and 7.10 examined, apparently for the first time, the behaviour of subjects in lifelike and abstract tasks possessing identical strategic structures.

7.9   Experiment II: Abstract and Lifelike Prisoner's Dilemma Games

A stringent test of the ecological validity of the PDG would require the comparison of behaviour in a conventional matrix version of the game to behaviour in a naturally occurring strategic interaction with an identical payoff structure. The experiment reported here (Colman, 1979a, Experiment II) was based on four structurally equivalent versions of the PDG, including a conventional payoff matrix and a simulation of a lifelike strategic interaction.

Two other treatment conditions were included in order to investigate the effects of monetary incentives, concerning which previous research (reviewed in Section 7.3) has generated equivocal results. In the positive incentive (PI) condition the subjects played for monetary gain, and in the negative incentive (NI) condition they played to conserve as much as possible of a monetary stake given to them at the start of the experiment. Although the payoffs in the NI condition were all zero or negative, the strategic choices facing the subjects were identical to the corresponding choices in the PI condition because the subjects stood to *save* the same amount in each outcome of the game as the PI subjects stood to *gain*. In the matrix (M) condition the subjects played for points, and in the lifelike simulation (S) they played for imaginary financial profits. An interval-scale equivalence was preserved between the payoff structures in all four treatment conditions.

It was hypothesized that the S condition would elicit fewer C choices than the other (abstract) conditions. This hypothesis was based on the fact that the S condition was a simulation of an economic conflict: two restaurateurs competing for the same clientele have to decide whether or not to provide expensive floor shows in their respective establishments. There is reason to suppose that this type of situation engages values encouraging competitiveness in our culture (Eiser and Bhavnani, 1974) and that subjects are therefore likely to manifest a degree of ruthlessness which might be inhibited in decision contexts lacking any specific contextual interpretation.

A difference between the choice behaviour of subjects in the PI and NI conditions was hypothesized on essentially intuitive grounds. Although they are strategically (indeed logically) identical, the problems facing the subjects in these two task situations are evidently quite unlike each other from a psychological point of view. It was felt that this was likely to manifest itself in differences in the subjects' behaviour in these two situations, but the nature of the differences, in the absence of previous evidence, was not predicted.

Method

The subjects were 80 undergraduate and postgraduate students, randomly assigned to treatment conditions in pairs. In the PI condition, the following tape-recorded instructions were played to the subjects:

> Your task will consist of making a series of 30 decisions. You may each earn up to 120 half pence (that is 60p) in this experiment if your decisions turn out well, but you may get less. How much money you eventually go home with will depend not only on the decisions which you make, but also on the decisions which the other person makes. You are advised, therefore, to consider each decision carefully. After each decision you will receive payment in the form of 1, 2, 3, or 4 half pence. In each case your decision, as shown on the cards in front of you, will be either R for Right or L for Left.
>
> The rules governing payment are as follows. If you both choose R, you will each receive 2 coins. If you both choose L, you will each receive 3 coins. But if one of you chooses R and the other chooses L, then the one who chooses R will receive 4 coins and the one who chooses L will receive 1 coin. I shall repeat these details; please feel free to make notes if you have difficulty memorizing them. [The previous three sentences were repeated.]
>
> Each decision will be made without knowledge of what the other person has chosen, by pointing to one of the two cards in front of you. Your decisions will be irreversible and any attempt to communicate with or indicate your feelings to the other person, for example by sighing or laughing, will force the experimenter to terminate the experiment. When you have both reached a decision the experimenter will announce the choices. . . . Your sole objective is to accumulate as many coins as possible. After 30 trials your money will be converted into more convenient coins to take home.

In the NI condition, the instructions were identical except for the necessary alterations to take account of the fact that the payoffs were zero or negative. Each subject received 120 coins to begin with, and the payoffs were derived from those of the PI condition by subtracting four units from each payoff. In both conditions, L choices correspond to the C strategy of the underlying PDG. The interval-scale equivalence of the two payoff structures is evident from the fact that in each structure $S + 1 = P$, $P + 1 = R$, and $R + 1 = T$.

Subjects in the M condition received the same instructions as those in the PI condition, except that "points" rather than "half pence" and "payments" were referred to, and the instructions were accompanied by the payoff matrix shown in Matrix 7.3.

In the S condition, the game was presented as a lifelike simulation of a decision making predicament facing two competing restaurateurs. The payoffs built into this simulation were designed so as to bear a ratio-scale equivalence to those used in the PI and M conditions. In all three cases T, R,

Matrix 7.3

Other's choice

		L	R
L		3,3	1,4
R		4,1	2,2

Your choice (L, R rows)

P, and *S* occur in the ratios 4:3:2:1. The payoffs in the S condition, like those used in the PI and M conditions, bear an interval-scale equivalence to the payoffs in the NI condition. The subjects in the S condition received the following tape recorded instructions:

> Your task will consist of making a series of 30 decisions. You may do very well for yourself if your decisions turn out well, but you may be less successful. The outcome in each case will depend on the decisions which the other person makes. You are advised, therefore, to consider each decision carefully.
>
> Your decisions will be based upon the following hypothetical situation. You have just taken over the ownership of a small restaurant which specializes in Greek food. There is another Greek restaurant just opened in the same town which is in every way similar to yours, and which caters for the same clientele. The problem facing you at the beginning of each week is whether to provide a (rather costly) floor show or not, and your sole consideration is to maximize profits. In each case your decision, as shown on the cards in front of you, will be SHOW or NO SHOW.
>
> The following facts are known to both of you from the experience of the previous owners: if you both provide floor shows, you will both make exactly the same standard profit during that week. If neither of you provides a floor show, your overheads will be less and so your profits will both be 50 per cent higher than the standard. If, however, one of the restaurants provides a floor show and the other does not, then the one which provides the floor show will attract business from the other and will make double the standard profit for the week, while the restaurant without a floor show will make half the standard profit. I shall repeat these details; please feel free to make notes if you have difficulty memorizing them. [This paragraph was repeated omitting the final sentence.]
>
> Each decision will be made without knowledge of what the other person has chosen [etc. as above]. Remember that your sole objective is to accumulate as much profit as possible. After 30 trials you will be able to see how you have done over the whole 30-week period.

The strategic structure of this simulated situation is identical to that of the abstract games. The NO SHOW choice corresponds to the *C* strategy of the underlying PDG, since it yields the lowest profit if the other chooses SHOW (corresponding to *D*). As in the matrix, the profit is twice as high if both choose SHOW, three times as high if both choose NO SHOW, and four times as high to a unilateral SHOW chooser.

After each trial the subjects rated their degree of satisfaction with the outcome of the trial on a five-point scale from "very displeased" to "very pleased". When 30 trials had been completed each subject was interviewed separately about his general strategy and attitude towards the other player before being debriefed.

Results

The main dependent variable was the number of *C* choices (L or NO SHOW) per pair of subjects. These were tabulated by treatment conditions and trial blocks. Each trial block represented 10 joint decisions, hence the maximum number of *C* choices in each trial block was 20. The means are shown in Table 7.2.

Table 7.2   Mean number of *C* choices per pair ($N = 40$)

Treatment condition	Trial block 1	Trial block 2	Trial block 3
PI	6.50	6.20	6.70
NI	8.80	6.10	7.00
M	6.40	5.70	6.20
S	4.10	3.40	3.10

The overall grand mean is 5.85, indicating that about 30 per cent of the choices across treatment conditions and trial blocks were *C* choices. This figure is rather low compared with previous PDG results, but it can be attributed largely to the extremely low frequencies of *C* choices in the S condition.

The effect of treatment condition was marginally significant ($F(3,36) = 2.65, 0.05 < p < 0.10$), and a planned comparison between the abstract conditions (PI, NI, and M) and the lifelike condition (S) confirmed that significantly fewer *C* choices were made in the latter ($p < 0.05$). The trial block and interaction effects were non-significant.

For each subject, a product–moment correlation coefficient was computed between payoffs and satisfaction ratings on corresponding trials. (The rationale for this was explained in the introduction to Experiment I.) The mean correlations in the four treatment conditions were as follows. PI: $r =$

0.627, NI: $r = 0.735$, M: $r = 0.602$, S: $r = 0.773$. The grand mean was $r = 0.685$, indicating that slightly less than half of the variance in satisfaction ratings is explained by the explicit payoffs according to which the games were structured. The differences between the mean correlations were not significant ($F(3,76) = 1.80$, n.s.), but it is noteworthy that the S condition yielded the highest and the M condition (the one most frequently used in gaming experiments) the lowest mean correlation.

A subsidiary analysis of the choices of 12 pairs of subjects both of whose members had exhibited extremely high payoff–satisfaction correlations (with a median of $r = 0.868$) was performed. Although statistically non-significant, the pattern of results was essentially the same as that found in the main analysis.

The comments made by subjects during the post-experimental interviews revealed that feelings tended occasionally to run high, and most subjects were extremely ego-involved in the tasks. This was particularly the case in the PI and S conditions, where some subjects described their (highly competitive) opponents as "despicable", "bastard", "selfish", "stubborn", etc. Other subjects described their opponents as "sensible", "intelligent", etc., but comments of this type were significantly less frequent in the S condition than in the abstract conditions ($z = 1.99$, $p < 0.05$).

Discussion

Since the four versions of the game were structurally equivalent, all significant differences can be attributed to strategically irrelevant features of the decision contexts in which the subjects found themselves. No significant incentive effects were found, which reinforces the suggestion in Section 7.3 that monetary payoffs do not influence behaviour in a pronounced or consistent way. The decline in frequency of C choices over trial blocks was statistically non-significant (although a Time Series Analysis, not reported here, confirmed previous findings of a marked decline in the M condition).

The major hypothesis was that the lifelike simulation would elicit fewer C choices than the abstract versions of the game and this hypothesis was confirmed. This result corroborates and extends Eiser and Bhavnani's (1974) finding that an "economic" decision context tends to enhance competitive behaviour in the PDG, presumably because of the cultural value associated with competitiveness in this sphere. It suggests, furthermore, that the ecological validity of abstract gaming experiments depends, in part at least, on psychological features of the naturally occurring interactions to which they are generalized.

The payoff–satisfaction correlations revealed, as in Experiment I, that few subjects adhered strictly to the payoff structure presented to them in spite of exhortations to the effect that their sole objectives should be to maximize their own points, monetary gains, or imaginary financial profits. Less than half of the variance in subjective preference ratings could be explained by the explicit payoffs. A striking finding was that subjects adhered most closely to the explicit payoff structure in the lifelike simulation and least closely in the conventional matrix condition. This appears to undermine one of the major arguments used to justify the use of "meaningless" payoff matrices in experimental gaming, namely the suppression of extraneous sources of utility. The evidence of this experiment suggests that more meaningful or lifelike decision tasks may be more appropriate for investigating strategic choices. In any event it is important to discover in what circumstances and to what extent the findings of conventional gaming experiments can be generalized to more lifelike situations. The results reported here suggest that they cannot be generalized straightforwardly in all circumstances.

7.10   Experiment III: Abstract and Lifelike Chicken Games

Sermat (1970) compared behaviour in the game of Chicken with behaviour in a picture-interpretation task and a "Paddle game", but the strategic structures of the tasks were not the same, they were not compared in a single experimental design with irrelevant variables controlled, and none of the tasks were truly lifelike (see Section 7.8). In the experiment described below (Colman, 1979a, Experiment III) a more rigorous examination of the generalizability of behaviour in a conventional matrix version of the Chicken game to a lifelike simulation of a structurally equivalent strategic interaction was undertaken. Apart from the use of a Chicken rather than a PDG payoff structure and a completely different lifelike simulation, the experimental design and procedure were identical to those described in Experiment II above.

The major hypothesis was that the lifelike simulation (S) condition would elicit fewer C choices from the subjects than the positive and negative incentive conditions (PI and NI) and the matrix condition (M) which were all abstract and "meaningless" in character. This hypothesis was based on the fact that the C choice in the game of Chicken is both the cooperative and the cautious strategy (see Section 6.5). The S condition involved the simulation of an economic conflict between two firms exporting fresh vegetables to an island by sea or air freight. A decision context of this type may be expected to engage cultural values associated with competitiveness and risk taking. The competitiveness assumption is supported by the findings of Eiser and Bhavnani (1974) and those of Experiment II. The cultural value associated with risk taking in economic and other contexts has been established by numerous experiments on the risky shift and the group polarization phenomenon (Myers and Lamm, 1976; Lamm and Myers, 1978).

Incentive effects were hypothesized on common-sense grounds although the evidence of previous research is equivocal (see Section 7.3). In particular, a greater frequency of *C* choices was hypothesized in the NI than in the PI condition since there seems to be less justification for taking risks (i.e. making *D* choices) when one is attempting to *save* money than when one strives to *acquire* money.

Method

The subjects were 80 undergraduate and postgraduate students at the University of Leicester, randomly assigned to the PI, NI, M, and S treatment conditions in pairs. The data from one pair of subjects in the NI condition were unfortunately mislaid; the missing scores from this treatment condition were estimated by the method of unweighted means.

Matrix 7.4

Other's choice

		L	R
	L	3,3	2,4
Your choice			
	R	4,2	1,1

The procedure was identical to Experiment II apart from the payoff structure used and the content of the lifelike simulation in the S condition. The payoff structure is shown in Matrix 7.4, in the form used in the M condition. In the PI condition the subjects played for half pence. In the NI condition they were given 120 half pence to begin with and received zero or negative payoffs as in Experiment II (four units were subtracted from each of the payoffs shown in Matrix 7.4). In the M condition the subjects played for points. In the S condition they played for imaginary financial profits according to the following instructions:

Your decisions will be based on the following hypothetical situation. You have just taken over ownership and management of a firm whose sole business involves supplying a small island with fresh vegetables. There is another firm just started business in the same area which is in every way similar to yours, and which sells its vegetables to the same island. The problem facing you each morning is whether to send your goods to the island by air or by sea, and your sole consideration is to maximize profits. In each case your decision, as shown on the cards in front of you, will be AIR or SEA.

The following facts are known to both of you from the experience of the previous owners: If you both use the air freight, the airport authorities will not despatch such a large cargo immediately, but will wait for the departure of a large aircraft in the afternoon, on which to send the goods. The goods will consequently arrive very late at the market, and each firm will in that case have to be content with a minimal standard profit on that day. If both firms use the sea crossing, the goods will arrive only slightly late, and in these circumstances each firm will make three times the standard minimal profit on that day. If, however, one of the firms uses the air freight and the other uses the sea freight, the firm which uses the air freight will have its (relatively small) cargo despatched immediately on a light aircraft, while the firm which uses the sea freight will again have its goods delivered slightly late. In that case, the firm using the air freight will make four times the standard minimal profit for the day (since his goods will be on sale from the opening of the market) while the firm using the sea freight will make only double the standard minimal profit (since by the time its goods are on sale many customers will have bought the goods supplied by the other firm).

Summary:—

Both send by air—each gets x;
Both send by sea—each gets $3x$;
One by air, one by sea—$4x$ and $2x$ respectively.

In this treatment condition, SEA corresponds to the L choice in the abstract conditions; these are of course the C strategies in the underlying Chicken game. An interval-scale equivalence exists between the payoff structures of the four treatment conditions, and a ratio-scale equivalence between the PI, M, and S conditions.

Results

The mean numbers of C choices per pair in each block of 10 trials are shown in Table 7.3. The overall grand mean is 8.09 (out of a possible total of 20 C choices per pair per trial block), indicating that approximately 40 per cent of the choices across conditions were C choices. The frequencies in all conditions are distinctly on the low side compared with previous findings on Chicken (e.g. Rapoport and Chammah, 1969) and may reflect a cultural difference between American and British students.

The only statistically significant effect is the one due to treatment conditions ($F(3,36) = 4.19, p < 0.025$). A planned comparison confirmed that significantly fewer C choices were made in the S condition than in the abstract (PI, NI, and M) conditions ($p < 0.05$), but the hypothesized difference

Table 7.3   Mean number of *C* choices per pair ($N = 40$)

Treatment condition	Trial block 1	Trial block 2	Trial block 3
PI	10.40	11.10	9.40
NI[a]	8.67	8.56	8.56
M	7.90	7.10	6.70
S	6.90	6.10	5.80

[a] Including unweighted means estimates for one pair.

between the PI and NI conditions was non-significant (and in the opposite direction to the hypothesis).

The mean product–moment correlations between payoffs and ratings of subjective satisfaction with outcomes (calculated separately for each subject) were as follows. PI: $r = 0.779$, NI: $r = 0.695$, M: $r = 0.654$, S: $r = 0.721$. The grand mean is $r = 0.712$, which indicates that slightly more than half of the variance in ratings of satisfaction is explained by the explicit payoffs in the games. As in Experiment II, the data suggest that subjects in the conventional matrix (M) condition deviated most markedly from the explicit payoff structure, but the differences between the mean correlations in the various treatment conditions are statistically non-significant.

A subsidiary analysis of the *C* choices of 12 pairs of subjects who had exhibited very high payoff–satisfaction correlations (with a median of $r = 0.863$) revealed a pattern of results essentially similar to that found in the main analysis, although the effect due to treatment conditions failed to reach significance with such a small N ($F(3,8) = 1.93$, $0.10 < p < 0.20$).

In the post-experimental interviews, most subjects gave evidence of emotional involvement in the game. A large number in all treatment conditions described their opponents as "selfish", "greedy", "inflexible", "stubborn", "obstinate", "uncooperative", etc., and epithets like "twat", "a bit of a capitalist" and "pig-headed" were also used. One subject in the S condition summed up the point of the experiment as follows: "I have never before done an experiment of this kind, but I feel it demonstrates the difficulties of making decisions which involve other people's judgments".

Discussion

The results of this experiment confirm those of Experiment II. Once again, no significant incentive effects were found, which lends further credence to the suggestion that monetary payments do not influence experimental gaming behaviour in any consistent or marked fashion. The post-experimental interviews gave no indication that subjects were more ego-involved in the incentive than in the non-incentive conditions. A significant difference was, however, found between frequencies of C choices in the abstract, "meaningless" decision contexts on the one hand and the lifelike simulation on the other. As hypothesized, significantly fewer C choices were made in the lifelike simulation, presumably because it involved an economic conflict between imaginary firms which was calculated to engage cultural values associated with competitiveness and risk taking.

As in Experiment II, the evidence of the payoff–satisfaction correlations suggested that subjects in the conventional matrix version of the game introduced extraneous sources of utility and departed from the explicit payoff structure to a larger degree than those in other treatment conditions. Although the correlation differences failed to reach statistical significance, they suggest, once again, that the justification in terms of controlling extraneous sources of utility for the use of abstract decision contexts in experimental gaming is unfounded.

The results of this experiment do not support the view that the findings of conventional experiments with the game of Chicken can be generalized in a straightforward fashion to more lifelike strategic interactions with this payoff structure. The evidence strongly suggests that strategically irrelevant psychological factors play a large part in determining the frequencies of C choices in Chicken-type interactions. This is the same conclusion that emerged from the investigation of abstract and lifelike Prisoner's Dilemma games in Experiment II.

7.11 Summary

This chapter opened with some general comments on two-person, mixed-motive experimental games and a "review of reviews" of the literature in this area. In Section 7.2 the effects of strategic structure on choice behaviour were discussed, and it was concluded that experimental subjects often conform closely to the predictions of informal game theory. In the Prisoner's Dilemma game, however, there is a tendency for subjects to become "locked in" to the unsatisfactory DD outcome when the game is repeated many times, and in the Maximizing Difference game apparently irrational choices are common, indicating that the explicit payoff structure does not accurately reflect the subjects' subjective preferences. Section 7.3 focused on the effects of payoff variations and monetary incentives, and Section 7.4 discussed decomposed Prisoner's Dilemma games, the "trucking" game, and the effects of verbal communication between players on strategic choices. Section 7.5 was devoted to experiments on responses to prearranged programmed strategies of the other player, including 100 per cent C, 0 per cent C, tit-for-tat, "reformed sinner", and "lapsed saint". In Section 7.6 the evidence that females frequently behave more "competitively" than males was evaluated and reinterpreted. Section 7.7 centred on the relationship between cooperative or competitive intentions and attributions of intentions to the other player. The evidence suggests that competitive people are more inclined than cooperative people to believe that others are similar to themselves. In Section 7.8 seven experiments designed to investigate the generalizability of two-person mixed-motive experimental games to more lifelike strategic interactions were critically reviewed, and in Sections 7.9 and 7.10 two new experiments devoted to this question were reported. Both of the new experiments led to the conclusion that strategic choices in lifelike situations differ systematically and predictably from those in abstract matrix games with equivalent payoff structures.

8

Multi-Person Games: Social Dilemmas

8.1 Multi-Person Game Theory

FROM a theoretical point of view, one-person games are simpler than any others: the solitary player is motivated simply to maximize his payoff, or his expected payoff if the game contains an element of risk. The involvement of a second player whose choices also affect the outcomes leads to new and qualitatively different problems. In a two-person game, strategic considerations which cannot be reduced to simple maximization problems have to be taken into account and, depending on the payoff structure, the players are motivated to cooperate, to compete, or to strike a suitable balance between cooperation and competition. An additional source of strategic complexity arises in games involving three or more players, which are called multi-person or N-person games. Georg Simmel, the turn-of-the-century German sociologist and philosopher, was apparently the first to draw attention to an important phenomenon which emerges when the transition is made from two-person to three-or-more-person groups: "The essential point is that within a dyad, there can be no majority which could outvote the individual. This majority, however, is made possible by the mere addition of a third member" (quoted in Luce and Raiffa, 1957, p. 155). In game theory terminology, the strategic implications of *coalitions* have to be taken into account in multi-person games, unless the interests of the players happen to coincide exactly.

Multi-person interdependent decisions in which the decision makers' interests coincide exactly are pure coordination games. In these rather special cases, the distinction between two-person and multi-person games is relatively unimportant, and examples of both kinds were discussed in Chapter 3. Since the players' preferences are identical, no coalition at loggerheads with any other can form; the only profitable coalition is the one containing all of the players. Deciding whether to dress formally or informally for a social gathering, if one's sole objective is to match the choice(s) of the other guest(s), presents a similar strategic problem whether the occasion is an intimate tête-à-tête or a large party. Coordination may be achieved by explicit communication between the guests or, if that is not feasible, by the discovery of socially or culturally "prominent" features of the situation which can serve as focal points for coordination (see Section 3.2).

Zero-sum and mixed-motive multi-person games, on the other hand, are qualitatively different from their two-person counterparts because only in multi-person games can the players be motivated to split into alliances, blocs, cliques, caucuses, unions, or cartels whose collective interests diverge. This possibility can radically alter the character of a game. A multi-person zero-sum game, for example, can no longer be considered to be strictly competitive if—as is often the case—some of the players can benefit by forming a cooperative coalition against the others.

Formal multi-person game theory falls into two categories. The *non-cooperative* approach is based on the normal form and an extension of the equilibrium point concept discussed at length in previous chapters. (In this context, the adjective *non-cooperative* refers to the rules of the game rather than to its payoff structure.) In a non-cooperative multi-person game the players choose their strategies independently; communication and bargaining are not allowed and coalition formation is therefore usually assumed to be impossible. Non-cooperative "solutions" to multi-person games are open to all of the objections raised against equilibrium "solutions" to two-person, mixed-motive games (see Chapter 6), but they enable interesting predictions to be made in certain cases. The *cooperative* approach is based on models in which binding agreements and coalition formation are allowed by the rules, as they are in many everyday multi-person interactions. Von Neumann and Morgenstern (1944) developed this approach at great length in their classic *Theory of Games and Economic Behavior* and a large body of mathematical theory has been built upon it. The cooperative "solutions" to multi-person games unfortunately tend to lack any clear operational meaning when applied to real strategic interactions and many of them are unclear even from a purely mathematical point of view. Jean-Pierre Séris (1974) has commented: "The diversity of theories . . . proves that if the players wish to arrive at a tacit agreement, even or perhaps *especially if they are mathematicians*, there is a good chance that they will not settle upon the same point" (p. 91, emphasis added).

A more fruitful approach is the intensive analysis of the properties of particular games which, for one reason or another, are of special interest. Equilibrium points, coalitions, minimax strategies, and other concepts of formal game theory are applied in a common-sense way in the informal analysis of specific games. During the 1970s two multi-person games with important implications for contemporary social problems began to attract the attention of game theorists and social scientists. The United States involvement in an escalating multilateral military conflict in Vietnam led to the discovery of the Dollar Auction game. This game models some of the dynamic features of escalation and the *Concorde fallacy* (the tendency to invest ever larger amounts in a project when one has "too much invested to quit"). At about the same time, public concern about inflation and voluntary wage restraint, multilateral disarmament, pollution, overpopulation, and conservation of energy and other scarce resources led to penetrating analyses of the *N*-Person Prisoner's Dilemma. These two games, particularly the latter, have aroused considerable interest among psychologists, sociologists, economists, and moral philosophers.

The account of formal multi-person theory in Sections 8.2, 8.3, and 8.5 is brief and selective and focuses only on the essential ideas. In Section 8.4 the process of coalition formation in a three-person strategic interaction is illustrated in Harold Pinter's play *The Caretaker*. Section 8.6 is devoted to the Dollar Auction game and the Concorde fallacy, and the theory of *N*-Person Prisoner's Dilemma is outlined in Section 8.7. A general theory of compound games is sketched in Section 8.8, and a brief summary of the chapter is given in Section 8.9.

8.2   Non-Cooperative Games: Equilibrium Points

Situations frequently arise in which several decision makers have to choose independently from among two or more alternative courses of action. Provided that the decision makers have well-defined preferences among the outcomes, and the outcomes depend on the choices of the decision makers and possibly also of Chance or Nature, the situation can be modelled by a multi-person game; a classic example is seven-handed poker. The normal form of such a game cannot be represented by a payoff matrix of the kind used to depict a two-person game because the number of players exceeds the number of spatial dimensions, but it can nevertheless be conceived as an abstract entity. This type of model is appropriate for multi-person interdependent decision making where communication is impossible and there are no prospects of coalition formation. Coalition formation is, of course, often impossible in everyday strategic interactions, and it is sometimes explicitly forbidden where it might otherwise be possible. The antitrust laws in the United States and the Monopolies and Restrictive Practices Act in Britain, for example, are expressly designed to ensure that certain economic games are played non-cooperatively.

The formal solutions to non-cooperative games are based on equilibrium points. An equilibrium point in a multi-person game is an outcome which gives none of the players any cause to regret their strategy choices when the choices of the others are revealed. In other words, no player can achieve a better payoff by deviating unilaterally from an equilibrium point. John Nash (1950, 1951) succeeded in generalizing the fundamental minimax theorem (see Chapter 4 and Appendix A) by proving that every finite game possesses at least one equilibrium point in pure or mixed strategies. Thus every finite multi-person game has at least one non-cooperative "solution" corresponding to a Nash equilibrium point; the problem is usually that it may have several irreconcilable "solutions" in this sense.

Any finite, two-person, zero-sum game can be convincingly solved by finding an equilibrium point, as shown in Chapter 4, and computational procedures are available for achieving this. If more than one equilibrium point exists, then (a) they are *equivalent* in the sense of yielding the same payoffs to the players so that one is as good as another from the points of view of both players, and (b) they are *interchangeable* in so far as any combination of strategies, each of which corresponds to one of the equilibrium points, is bound to produce an equilibrium outcome.

It was shown in Chapter 6 that these desirable qualities are not shared by equilibrium points in two-person mixed-motive games: the players may prefer different equilibria, and if each chooses a strategy corresponding to his preferred equilibrium the outcome may be a non-equilibrium. In multi-person games these problems are compounded. There are frequently a large number of non-equivalent and non-interchangeable Nash equilibrium points and, what is worse, no-one has managed to devise a computational procedure or algorithm for finding them. Even if all of the players are perfectly rational, the outcome of a multi-person game may be a non-equilibrium, and if it is repeated many times there may not be any tendency for the choices to converge towards a unique outcome. In certain specific cases, some of which are discussed in Sections 8.7 and 8.8 and Chapter 11, it is however possible to determine how rational players will behave.

8.3   Cooperative Games: Characteristic Functions

Let us now consider games in which the players are able to communicate with one another and form binding agreements or coalitions. A coalition will, of course, be attractive to players only if it offers its individual members more than they can get independently through non-cooperative play. Games in which this is the case for at least one possible coalition are called *essential*; those in which the players cannot improve their payoffs through cooperative coalition formation are *inessential*. Every two-person, zero-sum game is inessential because the minimum payoff that each player can guarantee for himself is determined by the value of the game and cannot be improved upon through collaboration; but the Prisoner's Dilemma game, for example, is essential because both players can benefit by establishing a two-person cooperative coalition. Zero-sum games involving more than two players, unlike their two-person counterparts, are often essential, and it is therefore misleading to describe them as strictly competitive.

Von Neumann and Morgenstern (1944) approached the analysis of multi-person cooperative games by introducing the idea of a *characteristic function* as a point of departure. The characteristic function of any game is simply a rule which assigns a value to every logically possible coalition that might form. This can be illustrated most simply for a two-person game such as the Prisoner's Dilemma shown in Matrix 8.1.

Matrix 8.1
Prisoner's Dilemma game

II

		C	D
I	C	3,3	1,4
	D	4,1	2,2

From a mathematical point of view, four coalitions are possible: the "coalition" of no players, the "coalition" consisting of Player I alone, the "coalition" consisting of Player II alone, and the coalition of both players. The first three of these "coalitions" are not coalitions in the everyday sense of the word—hence the quotes—but mathematically speaking a coalition is simply a subset of the set of players, and the one-element and empty subsets have to be included for the sake of completeness. The characteristic function assigns a value, denoted by $v(S)$, to each coalition. In this notation S denotes any subset of players and $v(S)$ is the value of the game to that subset; that is, the minimum payoff that the subset can guarantee to receive by coordinated action irrespective of the strategies chosen by the remaining players (if any).

Let us work out the characteristic function of the Prisoner's Dilemma game shown in Matrix 8.1. It is assumed that the value of the game to the empty coalition (designated by the symbol Φ) is always zero, hence the first value of the characteristic function is $v(\Phi) = 0$. Player I, acting alone, can do no better than guarantee that his payoff will be at least 2 units; this is Player I's *security level*, and it is ensured by the choice of the minimax strategy D. (In some games, of course, a player can ensure a better security level by using a mixed strategy, but that is not the case in this game.) Player II's value is also 2 units since the game is perfectly symmetrical; no pure or mixed strategy gives Player II a better guaranteed minimum. Two further values of the characteristic function have thus been found: $v(\text{I}) = 2$ and $v(\text{II}) = 2$. Finally, if Players I and II form a grand coalition acting as a single player, agreeing to choose row C and column C respectively, then the two-person coalition is guaranteed a total payoff of 6 units. This is the best the grand coalition can achieve by concerted action; CD, DC, and DD all pay less to the coalition as a whole. It is clear, therefore, that the final value is $v(\text{I, II}) = 6$. The complete characteristic function of the game is as follows:

$$v(\Phi) = 0, \ v(\text{I}) = 2, \ v(\text{II}) = 2, \ v(\text{I, II}) = 6.$$

This method of analysis can be extended without difficulty to multi-person games. Consider the following three-person example. A bequest of 24 gold watches has been made by a benefactor to a corporation consisting of three shareholders. The shareholders have to reach a collective decision about how to divide the watches among themselves. The decision is made by majority voting, but the weight of each shareholder's vote is proportional to the percentage of shares owned by this person. (This type of arrangement is called a *weighted majority game*. It is common not only in business corporations but also in legislatures and trade unions where each vote is weighted in proportion to the number of members it represents.)

Suppose that Players I, II, and III own 50, 40, and 10 per cent shares respectively. Each can vote in favour of any feasible apportionment of the 24 watches among the three contenders; it is unnecessary to list the numerous options from which each player must choose. If an absolute majority in favour of some particular apportionment does not emerge from the voting—if there is a tie—then none of the players receives anything (the watches are donated to charity). We assume as usual that each player seeks solely to maximize individual gain. It is clear that none of the players acting alone can guarantee an outcome in which he receives any of the watches since no single shareholder has an absolute majority of the vote. Hence $v(\text{I}) = v(\text{II}) = v(\text{III}) = 0$. Players I and II jointly command a majority (they control $50 + 40 = 90$ per cent of the vote), and they can therefore form a coalition taking all of the watches for themselves: $v(\text{I, II}) = 24$. Players I and III also have a majority, but II and III do not. The value of the game to the grand coalition of all three players is obviously 24. The characteristic function of this weighted majority game is therefore:

$$v(\Phi) = 0;$$
$$v(\text{I}) = 0, \ v(\text{II}) = 0, \ v(\text{III}) = 0;$$
$$v(\text{I, II}) = 24, \ v(\text{I, III}) = 24, \ v(\text{II, III}) = 0;$$
$$v(\text{I, II, III}) = 24.$$

According to game theory, *one* of the winning coalitions, (I, II), (I, III), or (I, II, III) will form if the players are rational, but the characteristic function does not, on its own, generate any prediction about which *particular* coalition is most likely.

A large number of theories have been developed to make more precise prescriptions about coalition formation. Riker's (1962) *minimal winning coalition theory* is based on the idea that if a coalition is large enough to win, it should avoid taking in any superfluous members since the new members will demand a share in the payoff. If the players seek to maximize their individual payoffs, therefore, one of the smallest (minimal) winning coalitions should form. In the shareholders' game, this theory does not prescribe which of the two minimal winning coalitions (I, II) or (I, III), should form, but it rules out the non-minimal winning coalition (I, II, III).

Gamson's (1961) *minimum resource theory* is more specific. It assumes that each member of a winning coalition should demand a share in the payoff corresponding to his relative voting strength in the coalition. According to this principle, Player I can expect a larger individual share in the (I, III) coalition, where he has a 50 per cent vote compared with his partner's 10 per cent, than in the (I, II) and (I, II, III) coalitions where his relative voting strength is less. Minimum resource theory therefore predicts that Player I should prefer to form a coalition with Player III, and the winning coalition (I, III) should thus emerge. This coalition has the property of commanding the minimum resource—in this case the minimum total percentage vote—of any winning coalition. It commands a total of 60 per cent (Player I's 50 per cent plus Player III's 10 per cent), while the other winning coalitions command 90 and 100 per cent respectively.

The von Neumann–Morgenstern "solution" is based on the idea of *imputations*. An imputation is an apportionment of the payoff among the players which satisfies criteria of collective and individual rationality: the players receive jointly what they can get as a grand coalition (nothing is wasted) and each individual player gets at least as much as he can guarantee by acting independently. One imputation is said to dominate another if there is a subset of players who all prefer the first to the second and can enforce it by forming a coalition; they have both the will and the power to impose the dominant imputation. The *core* of a game is the set of all undominated imputations. If an imputation in the core exists, then it is stable: by definition, no player or subset of players is motivated to deviate from it to an imputation outside the core. Unfortunately, however, many games, including all multiperson, zero-sum games, have empty cores, in other words there are no undominated imputations. Von Neumann and Morgenstern therefore proposed the *set* of imputations with the following properties as a "solution" to a cooperative multi-person game: (a) no imputation in the set dominates any other in the set, and (b) every imputation outside the set is dominated by at least one imputation in the set. This set of imputations, considered in its totality, has a certain inner stability since no rational player or subset of players has both the will and the power to impose an outcome that is not part of the von Neumann–Morgenstern "solution".

There are two serious drawbacks to this "solution", one relating to uniqueness and the other to existence. In the first place, the von Neumann–Morgenstern "solution" set typically contains a plethora—often an infinity—

of imputations (they are interpreted as "standards of behaviour" governed by social and moral conventions) and there are no rational criteria for choosing among them. This is the case, for example, in the shareholders game. Concerning existence, von Neumann and Morgenstern (1944) had this to say:

> There can be, of course, no concessions as regards existence. If it should turn out that our requirements concerning a solution S are, in any special case, unfulfillable,—this would certainly necessitate a fundamental change in the theory (p. 42).

Twenty-four years later, William F. Lucas (1968) came up with a 10-person game which has no von Neumann–Morgenstern "solution".

8.4   Harold Pinter's "The Caretaker"

The Caretaker (Pinter, 1960) has been subjected to metagame analysis by Howard (1971, pp. 140–6). In this section the play is used for descriptive purposes only, to illustrate some of the principles outlined in the previous section.

There are just three characters in the play: Davies is a bigoted old tramp; Aston is a young man suffering from the after-effects of electric shock treatment; and Mick is Aston's younger brother. The action takes place in a room in a dilapidated London house owned by Mick, in which Aston is living.

In Act I Aston invites Davies, who needs to get himself "sorted out", to stay with him on a temporary basis. In Act II Mick visits the house while Aston is out and bullies Davies, who he says is "stinking the place out". When he discovers that Aston has invited the tramp to stay, he abruptly changes his tune and ingratiates himself with Davies. He tells Davies that he is the legal landlord and agrees to allow Davies to stay on as "caretaker".

In Act III Davies's attitude towards Aston cools dramatically. When Aston buys him a new pair of shoes, he complains that "they don't fit" although "they can do, anyway, until I get another pair". He says that Aston should go back to where he came from (the mental hospital) because he is "queer in the head", and he even menaces Aston with a knife. Aston eventually says "I think it's about time you found somewhere else. I don't think we're hitting it off". When Mick hears about these developments he shouts at Davies: "You're violent, you're erratic, you're just completely unpredictable. You're nothing but a wild animal, when you come down to it. . . . It's all most regrettable but it looks as though I'm compelled to pay you off for your caretaking work." Mick and Aston look at each other, smiling faintly.

After Mick has left the room, Davies makes an attempt to regain Aston's favour. He tries to convey his willingness to switch his allegiance from Mick to Aston: "I'd look after the place for you, for you, like, not for the other

. . . not for . . . for your brother, you see, not for him, for you, I'll be your man." When Aston reiterates his request for Davies to leave, Davies says "But you don't understand my meaning!", and the play ends with this attempted negotiation unresolved.

Three two-person coalitions are possible between the protagonists in this play. Each one forms, achieves its ends, and dissolves in one of the acts. In Act I the (Aston, Davies) coalition forms. It is beneficial to both since Aston gets a room-mate who treats him as "normal" and Davies gets a place to live. This coalition is unstable, however, because both Mick and Davies prefer the (Mick, Davies) coalition and, being in a majority, have the power to impose it. The (Mick, Davies) coalition, which forms in Act II, is better for Mick, who prefers Davies to give allegiance to him rather than to his brother, and it seems better for Davies because Mick is the legal landlord.

But the (Mick, Davies) coalition is also unstable. Both Mick and Aston can improve by forming a coalition against Davies. From a strategic point of view, Davies ought to allow the grand coalition (Aston, Mick, Davies) to form. But in Act III he commits the blunder of opting for what he takes to be a minimal winning coalition between himself and Mick, which has the power to "freeze" Aston out. He therefore turns on Aston. But both brothers now prefer Davies to leave; they form the (Aston, Mick) coalition and Davies is out in the cold.

Davies is quick to realize that the (Aston, Mick) coalition is also unstable: both he and Aston would benefit by re-establishing the original (Aston, Davies) coalition, and they have the power to impose it. Possibly because he feels wounded by Davies's past behaviour, Aston is unwilling to fall in with this plan. It is Davies who is coldly logical about the strategic structure of the situation: "But you don't understand my meaning!". The curtain goes down with the possibility of the whole cycle of coalition formation repeating itself. The problem arises from the fact that every outcome is dominated by another, so the core of the game is empty; in other words, for every coalition that might form there is a subset of players who have the will and the power to subvert it.

Harold Pinter (1972) has described his literary style as follows: "For me everything has to do with shape, structure, and overall unity" (p. 33). In *The Caretaker* he has depicted, with astonishing simplicity and clarity, the shifting pattern of coalitions in a three-person cooperative game with an empty core. Three coalition patterns among three players are represented in three acts. There is, of course, much more to the play than the strategic instability that it illustrates, but it is easy to miss the wood for the trees. Trussler (1973) has pointed out that, on seeing a Pinter play for the first time, "One can be merely baffled at a relatively simple but promising piece of work . . . or one can be deceived by the apparent profundity" (p. 182). From the point of view of game theory, *The Caretaker* is both simple and profound.

8.5 The Shapley Value

A large number of formal "solutions" to cooperative multi-person games, apart from the von Neumann–Morgenstern "solution", have been based on the characteristic function outlined in Section 8.3. None is entirely persuasive as a criterion of rationality, but one which has some interesting applications is the *Shapley value* (Shapley, 1953) outlined in this section. The Shapley value suggests a rational way of dividing a payoff among the players according to their relative *a priori* power.

To each coalition that might form in a cooperative game there is a value $v(S)$ defined by the characteristic function. The value to the empty coalition is zero by definition, so $v(\Phi) = 0$. If we imagine the players joining this initially empty coalition one by one until the grand coalition has formed, it is possible to determine how much value each player brings to the coalition by joining it. The Shapley value assigns to each player a share in the payoff proportional to the amount that the player adds to the coalition's value. Since the amount that a player adds may depend on the order in which the coalition builds up, all possible permutations are examined and it is the average amount of value that a player brings to the coalition, taking all possible orders into account, that determines his share in the payoff.

The Shapley value may be illustrated in terms of the shareholders' game introduced in Section 8.3. In this three-person weighted majority game, the voting strengths of the players are 50 per cent, 40 per cent, and 10 per cent respectively. Now imagine a grand coalition building up from scratch. Each of the players stands a chance of being the pivotal voter who changes an impotent coalition (with a guaranteed minimum payoff of zero) into a winning one (with a guaranteed minimum payoff of 24 gold watches). We must consider all possible orders of coalition formation.

The first player to join the coalition is never pivotal in this game because no single shareholder commands a majority vote (more than 50 per cent). If Player I, with a 50 per cent vote, joins first, then the second player to join is bound to be pivotal. If, on the other hand, Player II, with a 40 per cent vote, joins first and is subsequently joined by Player III, with a 10 per cent vote, an absolute majority is still lacking; Player I is pivotal in making the coalition winning by joining last. Continuing this analysis for all permutations of coalition formation, we arrive at the results shown in Table 8.1.

Table 8.1 Coalitions and pivots in the shareholders' game

Order	I–II–III	I–III–II	II–I–III	II–III–I	III–I–II	III–II–I
Pivot	II	III	I	I	I	I

In four cases out of six, Player I is pivotal in transforming a coalition into a winning majority. Players II and III are each pivotal in one of the six cases. It follows that if the order of coalition formation were purely random, the probability that Player I would be pivotal is 2/3, the probability that II would be pivotal is 1/6, and the probability that III would be pivotal is 1/6. If the payoff is distributed according to the Shapley value, each player getting the average amount that he adds to the coalition's value, then Player I gets 16 of the 24 gold watches, and Players II and III get 4 each.

This result is rather unexpected: it illustrates how the players' voting strengths can be a misleading index of their real *a priori* power in a weighted majority game. Player I commands only a 50 per cent vote, but his relative power is 67 per cent. The reason for this is that no winning coalition can form without Player I's cooperation. Even more surprising is the fact that Players II and III have equal *a priori* power. Although II commands 40 per cent of the vote and III only 10 per cent, the power index of each is approximately 17 per cent. This is because they are equally likely to be pivotal, and from Player I's point of view Player III is just as good a coalition partner as Player II in spite of the difference in voting strengths.

Applications of the Shapley value to measure the distribution of power in real-world decision making bodies have produced interesting results. Shapley and Shubik (1954) performed an analysis, essentially similar to the one described above, to the United Nations Security Council. They showed that the permanent members of the Security Council (the United States, the Soviet Union, China, the United Kingdom, and France), which have the power to veto resolutions, controlled 98.7 per cent of the power in the 1950s, while the six non-permanent members controlled only 1.3 per cent. In 1965 four new non-permanent members were added to the Security Council with the express purpose of reducing the relative power of the permanent members. The effect of this was, however, only marginal. The permanent members presently command 98.1 per cent of the power according to the Shapley–Shubik index compared with 1.9 per cent controlled by the 10 non-permanent members (Riker and Ordeshook, 1973, p. 172; Brams, 1975, p. 187).

Shapley and Shubik (1954) also calculated that, in the 533-member United States legislature, the power indices for a single congressman, a single senator, and the President are in the proportions 2:9:350. The President thus has 175 times as much *a priori* power as a representative and almost 40 times as much as a senator. The Shapley value has been applied, with results as instructive as these, to the United States Supreme Court (Schubert, 1959), the United States Electoral College (Mann and Shapley, 1964), and the Canadian legislature under a proposed constitutional amendment (Miller, 1973).

8.6   The Dollar Auction Game and the Concorde Fallacy

During the 1960s the United States became embroiled in an escalating military conflict in Vietnam. When the early victory that was promised did not materialize, one of the arguments against disengagement was that the United States had "too much invested to quit". Were they to withdraw before defeating the National Liberation Front (Viet Cong) and the North Vietnamese forces, their past investment in money and lives would be wasted. Successive administrations therefore intensified the conflict, only to find themselves increasingly overcommitted, and the argument for escalation in order to justify past investment, for what it was worth, grew increasingly plausible. When it became apparent that victory was beyond its grasp, the United States was eventually forced to withdraw without having achieved any of its objectives.

On a smaller scale and with less devastating consequences, the following European experience illustrates some of the same strategic properties of entrapment and escalation. The cost of the Anglo-French supersonic airliner, the Concorde, rose steeply during its development phase and it was soon apparent that the project was uneconomical. But British and French governments became increasingly strongly committed to it as its cost continued to rise. Continuing to invest in a project simply because so much has already been spent on it—instead of cutting one's losses by withdrawing— has been called the *Concorde fallacy* (Dawkins and Carlisle, 1976).

The dynamic properties of escalation and the Concorde fallacy are present in many areas of behaviour. Gamblers often feel impelled to throw good money after bad in an attempt to rid themselves of escalating debts, and marriage partners often find themselves trapped in escalating spirals of hostility and counter-hostility from which they feel increasingly incapable of extricating themselves on account of past emotional investments. But it was the Vietnam war that was mainly responsible for drawing the attention of game theorists to the strategic character of escalation. In 1971 Martin Shubik devised a simple multi-person game, the Dollar Auction game, which brings some of the features of strategic escalation and the Concorde fallacy into sharp focus.

The rules of the Dollar Auction game are as follows. An auctioneer auctions a dollar bill to the highest bidder. (Other currencies can, of course, be used.) The usual conventions of auctions are observed except for one special rule: *both* the highest bidder (who gets the dollar bill) *and* the second-highest bidder (who gets nothing) must pay the auctioneer amounts corresponding to their last bids. Suppose the bidding stops at 40 cents and the second-highest bid, just before the end, was 35 cents. In this case the dollar goes to the person who bid 40 cents, and the auctioneer collects 40 cents from this person and 35 cents from the second-highest bidder (who receives nothing in return). In order to ensure that the game is finite, an upper limit on bidding can be set; this ensures that the game does not continue forever. Another useful rule is that bids must be made in whole numbers of cents.

There are three important "moments of truth" in the Dollar Auction game. The first occurs when at least two bids have been made and a pause occurs. Suppose Player I bids 10 cents and Player II bids 20 cents. If no other player seems willing to raise the bidding, Player I faces the following dilemma: if he stands pat he is certain to lose 10 cents, but if he bids 30 cents, say, he stands a chance of winning 70 cents (if no further bids are made). The same dilemma, with modifications of the gains and losses, faces the second-highest bidder at every stage in the game.

The second "moment of truth" occurs if the bidding reaches 50 cents. The second-highest bidder must then decide whether to raise the bidding to 51 cents or more. This is a critical threshold from the auctioneer's point of view, since the sum of the two top bids exceeds 1 dollar and the auctioneer is bound to make a profit from that point on.

The most testing "moment of truth" occurs if the bidding reaches the 1-dollar mark. The second-highest bidder must now decide whether or not to bid more than a dollar for the 1-dollar payoff. If his last bid was 95 cents, and if no one else is willing to bid, he is bound to lose 95 cents if he stands pat. But if he bids 105 cents, say, he will lose only 5 cents provided that the bidding stops there. The danger exists, however, that the other bidder will go even higher for exactly the same reason, and the problem will present itself in an even more unattractive form. Once the bidding passes the 1-dollar point, the players are motivated to minimize their losses rather than to maximize their gains.

A player who has made a single bid in the Dollar Auction game can avoid loss only by dropping out before the number of active participants reduces to two, or by being the highest bidder when the bidding stops. Once two or more players have entered the bidding, it is certain that at least one of them will lose money. When the bidding passes the 1-dollar mark, two players are bound to lose, but each is still motivated to be the highest bidder in order to minimize the loss.

If communication among the bidders and coalition formation are forbidden (as they are in most auctions), the game can be analysed as a multi-person non-cooperative game. The non-cooperative "solutions", which do not seem to have been worked out before, are highly instructive. One possible strategy for a player may be expressed simply as: "Do not bid". Ignoring certain inessential complications, other strategies take the form: "Enter the bidding and continue (if necessary) until the highest bid is k cents" where k is a number between 1 and the upper limit (if any). A plausible strategy, for example, is to set $k = 100$, in other words to be willing to continue bidding until the highest bid is 100 cents. If there is an upper limit of 5 dollars, say, then each player has 501 pure strategies in the normal form of the game. If there is no limit, the game is infinite.

In this simplified model the equilibrium points can be characterized quite straightforwardly. They are the outcomes in which exactly one player selects some $k \geq 99$, i.e. chooses to bid as far as 99 cents or more, and every other player chooses "Do not bid". With an upper limit of 5 dollars, there are $402n$ pure-strategy equilibrium points, where n is the number of players. (Each of the n players has $501 - 99$ strategies which lead to equilibrium outcomes when all of the others choose "Do not bid".) If there is no limit, there is an infinite number of equilibrium points of the same type.

If the outcome is an equilibrium, the bidder will receive the dollar for the price of his first bid and will be more than satisfied with the outcome. The other players, provided that they are fully informed about the bidder's strategy, will have no cause for regret since they could not have won any money by entering the bidding, given the existence of a player willing to bid as far as 99 cents or more. In other words, no one can benefit by deviating unilaterally from this outcome.

All other combinations of strategies lead to non-equilibrium outcomes. To see why this is so, it is necessary to consider three general cases. First, suppose two or more players choose to enter the bidding. One of them is bound to end up as the second-highest bidder and to lose money; this player will have cause to regret his strategy choice since he could have avoided loss by not bidding at all, and the outcome is therefore a non-equilibrium. Second, imagine that exactly one player chooses to bid, but only as far as $k = 98$ cents or less. In these outcomes each of the other players, when all of the strategies are revealed, will regret not having chosen to bid as far as 99 cents with a gain of at least 1 cent. Finally, suppose every player chooses "Do not bid". In that case every player will regret not having been the sole player to bid 1 cent, thereby gaining 99 cents.

The minimax strategy for each player is "Do not bid"; it is the only way of guaranteeing the avoidance of loss. But if every player chooses minimax, each will have cause to regret not having been more enterprising. The minimax strategies, in other words, do not intersect in an equilibrium point. The game possesses numerous equilibrium points, but they are all non-symmetric and there are no rational criteria for choosing among them. These non-cooperative "solutions" are characteristically unconvincing, but they do suggest profitable ways of bending the rules of the game by issuing threats.

Suppose one player announces, before the game is played, that he is firmly committed to bidding beyond 99 cents if necessary, and to prove it he shows the other players an affidavit to this effect and a document instructing his lawyer to destroy his life's savings if he breaks this oath. If the other players are rational, he is bound to be the only bidder and he can win 99 cents. If two or more players issue similar threats, however, at least one of them will lose. Something of the sort happened in Vietnam.

Let us now examine the Dollar Auction game as a cooperative multi-person game in which communication and coalition formation are permitted by the rules. Shubik (1971) analysed the characteristic function of the game under the assumption that the auctioneer is one of the players and the payoff structure is zero-sum. In view of the fact that the auctioneer makes no choices that affect the outcomes, however, it seems more accurate to consider it a mixed-motive game in which the only players are the potential bidders. If there are exactly three players, I, II, and III, the characteristic function is as follows:

$$v(\Phi) = 0;$$
$$v(I) = 0, \quad v(II) = 0, \quad v(III) = 0;$$
$$v(I, II) = 0, \quad v(I, III) = 0, \quad v(II, III) = 0;$$
$$v(I, II, III) = 99.$$

The general principle is the same with any number of players. The grand coalition of all players acting in concert can always gain 99 cents; they achieve this by agreeing that one of them, nominated in advance, will bid 1 cent and no other bids will be made. This guarantees a gain to the coalition of 1 dollar for the price of 1 cent, and they can then divide the payoff of 99 cents among themselves. (The Shapley value assigns each player an equal share in the payoff since they have equal *a priori* power in the game.) No other coalition can guarantee to win anything.

Chapter 9 contains a review of empirical research based on the Dollar Auction game. To anticipate slightly, experimenters have found that the bidding invariably passes the 1-dollar mark, and experimental subjects (like American Presidents and European Ministers of Transport) become emotionally involved in the tragic spiral of escalation.

8.7   Multi-Person Prisoner's Dilemma

The Multi-Person or N-Person Prisoner's Dilemma (NPD), also known as the Commons Dilemma Game, was "discovered" simultaneously and independently by Schelling (1973), Hamburger (1973), and Dawes (1973). Although its existence had been hinted at before, these game theorists were the first to provide rigorous definitions and theoretical analyses of the NPD, and their work stimulated the rapid growth of a new field of experimental gaming (reviewed in Chapter 9). The abrupt emergence of the NPD was strongly suggestive of an idea whose time had come.

Without delving too deeply into the history of ideas, it is possible to pinpoint some of the major reasons for the explosion of interest in the NPD. The first reason is negative in character: after 20 years of intensive experimental investigation, the (two-person) Prisoner's Dilemma game (PDG) began to lose some of its lustre during the early 1970s for a number of reasons outlined in Chapter 7. Secondly, several social and economic problems with certain common strategic properties began to fill newspapers and television screens in the United States and Europe at about the same time. The most striking of these were the problems of inflation and voluntary wage restraint, the "energy crisis" and other problems related to the conservation of scarce resources, and the problems of environmental pollution. In addition, an increase in international tension following the Soviet invasion of Czechoslovakia, coupled with a proliferation of nuclear weapons, raised the problem of multilateral disarmament with renewed urgency. All of these problems—and many others besides—can be modelled by the NPD. The NPD corresponds to a truly remarkable range of real-world social problems, but a few simple examples will suffice.

(a) *The "invisible hand" and voluntary wage restraint*

In one of the most influential books in the history of economics, *The Wealth of Nations*, Adam Smith (1776) introduced the theory of the "invisible hand" as follows:

> It is not from the benevolence of the butcher, the brewer, or the baker, that we expect our dinner, but from their regard to their own self-interest. We address ourselves, not to their humanity but to their self-love, and never talk to them of our own necessities but of their advantages [p. 16]. . . . It is his own advantage, indeed, and not that of society, which he has in view. But the study of his own advantage naturally, or rather necessarily leads him to prefer that employment which is most advantageous to the society [p. 419]. . . . He generally, indeed, neither intends to promote the public interest, nor knows how much he is promoting it. . . . He intends only his own gain, and he is in this, as in many other cases, led by an invisible hand to promote an end which was no part of his intention. . . . By pursuing his own interest he frequently promotes that of the society more effectually than when he really intends to promote it [p. 421].

The theory of the "invisible hand" has been a cornerstone of conservative economic thinking for more than two centuries. Let us examine it in the context of collective wage bargaining.

It is in the individual self-interest of every trade union to negotiate wage settlements that exceed the rate of inflation in the economy as a whole. This remains true irrespective of whether other unions do the same or exercise restraint in their wage demands. But if all pursue their individual self-interest in this way, the prices of goods and services go up and everyone is worse off than if all exercise restraint. If each worker "intends only his own gain" he is led, not by an "invisible hand" which promotes the interest of society as a whole, but by an invisible claw which threatens to tear the collective fabric of the economy apart.

In October 1974 the Manifesto of the British Labour Party contained an outline of a "social contract" which was supposed to encourage trade unions to exercise voluntary wage restraint in order to reduce the rate of inflation. The social contract was designed to encourage collective rationality in wage bargaining in place of individual rationality:

> Naturally the trade unions see their clearest loyalty to their own members. But the social contract is their free acknowledgement that they have other loyalties—to the members of other unions too . . . to the community a a whole (quoted in Collard, 1978, p. 73).

In the event, the social contract was not a success. No doubt the major reason for its failure was that it left the strategic structure of the wage bargaining game intact.

(b) *Conservation of scarce resources*

There was a severe drought in Britain in the summer of 1976, and water became a scarce resource in most parts of the country. The mass media exhorted everyone to exercise restraint in their consumption of water—to place bricks in their lavatory cisterns, to avoid washing their cars and watering their gardens, and even to try showering with a friend. Government propaganda was based on a slogan used to encourage people to save energy when the price of oil doubled 2 years earlier: "Save it!".

The dilemma confronting an individual citizen during the drought or the "energy crisis" may be crystallized as follows. An individual benefits from restraint only if many thousands of others also exercise restraint. But, in that case, restraint on the part of a specified individual is clearly unnecessary: the benefit is reaped without the cooperation of this single individual, who nonetheless shares the reward. If, on the other hand, most of the others ignore the call for restraint, then restraint on the part of the individual is futile. From an individual's point of view, saving it—water or energy—is therefore either pointless or futile. It is evidently in each individual's rational self-interest to ignore the call for restraint irrespective of the choices of the others. But—and this is the rub—if everyone pursues individual rationality in this way, they are all worse off than if everyone is motivated by collective rationality and exercises restraint. If everyone tries to be a "free rider" then no-one gets a ride at all.

During the summer of 1976, only a small minority of British citizens heeded the call for water conservation (Hollis, 1979, p. 2). About 40 per cent made only token economies and 50 per cent made none at all. Edgeworth's (1881) dictum that "the first principle of economics is that every agent is actuated only by self-interest" (p. 16) proved roughly correct in this case as a *descriptive* principle, but as a *prescription* for rational action it is clearly deficient, as the conservation problem shows.

(c) *The tragedy of the commons*

This example is based on a phenomenon commented upon by Lloyd (1833) in a lecture on "the checks to population" and discussed more recently in a paper by Hardin (1968) which has acquired the status of a modern classic.

The problem is this. There are six farmers who have access to a common pasture on which to graze their cows. Each owns a single cow weighing 1000 lb. The pasture can sustain a maximum of six cows without deterioration through overgrazing: for every additional cow that is added to it, the weight of every animal decreases by 100 lb. This is important to the farmers because the weight of a cow is proportional to its value. Suppose each farmer has the opportunity to add another cow to the pasture. If one farmer decides to add a cow, his personal wealth in livestock will increase since he will then own two 900-lb cows instead of one 1000-lb cow. The same argument (with different figures) applies irrespective of how many of the other farmers decide to add a second cow; it is always in a farmer's individual self-interest to add one. But if all of the farmers do this, then each ends up worse off (with two 400-lb cows) than if they all stick to the status quo (with one 1000-lb cow each). The impoverishment of small farmers in England during the period of the enclosures in the eighteenth century may have been exacerbated by this phenomenon.

The three social dilemmas outlined above share a common underlying strategic structure. They can each be modelled by a multi-person game defined by the following properties:

(i) each player faces a choice between two options which may be labelled C (cooperate) and D (defect);

(ii) the D option is dominant for each player, i.e. each is better off choosing D than C no matter how many of the other players choose C;

(iii) the dominant D strategies intersect in a deficient equilibrium. In particular, the outcome if all players choose their non-dominant C strategies is preferable from every player's point of view to the one in which everyone chooses D, but no one is motivated to deviate unilaterally from D.

A game which satisfies these three conditions is by definition an NPD.

It is immediately obvious that the (two-person) Prisoner's Dilemma game

(PDG) defined in Section 6.6 is a special case of the generalized NPD defined above, for it satisfies all three conditions. In its n-person form it is surprisingly common in everyday strategic interactions. The D strategy corresponds to negotiating for a wage settlement in excess of the rate of inflation in Example (a) above, disregarding the call for water or energy conservation in (b), and adding a second cow to the common in (c). Here are a few more examples of D choices: carrying a concealed weapon or (at a national level) manufacturing nuclear weapons (Schelling, 1973); refusing to join a union (Messick, 1973); rushing for an exit during an escape panic (Dawes, 1975); standing on tiptoe to watch a parade (Caldwell, 1976); neglecting to contribute towards the cost of a neighbourhood playground (Hamburger, 1979, chap. 7); and increasing the size of one's family in India (Dawes, 1980). In each case—and in numerous others that have been discovered—it is in an individual's self-interest to choose the indicated D option regardless of the choices of the others, but everyone is better off if they all choose the alternative C option. Adam Smith's theory of the "invisible hand", in its broadest interpretation, is refuted in every case. What is clearly required is some form of collective rather than individual rationality, such as the one embodied in Kant's categorical imperative discussed in Section 6.6. There is room for N martyrs in a society of N individuals, and virtue is its own reward in such a utopia.

Let us now examine the formal structure of the NPD in more detail. The simplest instance of it involves three players and integral payoffs of 1, 2, 3, and 4 units. Two assumptions that simplify the analysis further are that the game is *symmetric*, that is to say it is the same from every player's point of view, and that each player's payoff function is *linear*, that is, his payoff from a C or D choice is directly proportional to the number of players choosing C. If the degree to which a D choice is preferable to a C choice, from an individual point of view, is constant across all possible contingencies, then the payoff structure is as shown in Matrix 8.2.

Matrix 8.2   Three-Person Prisoner's Dilemma

Number Choosing C	Number Choosing D	Payoff to Each C-Chooser	Payoff to Each D-Chooser
3	0	3	—
2	1	2	4
1	2	1	3
0	3	—	2

As shown in the first row of the matrix, the payoff to each C-chooser is three units if all three players choose C, and in this case the payoff to the (non-existent) D-chooser is undefined, hence the dash in the last column. The second row shows what happens if two players choose C and the remaining player chooses D; in that case the payoff to each C-chooser is two units and the payoff to the solitary D-chooser is four units, and so on. The game is an NPD because it satisfies the three defining conditions mentioned earlier: (i) each player chooses between two options; (ii) a D choice pays better than a C choice no matter how many of the *other* players choose C, so that the D choice is dominant; and (iii) each player is better off when everyone chooses C than when everyone chooses D, but no-one is motivated to deviate unilaterally from D, in other words the dominant strategies intersect in a deficient equilibrium.

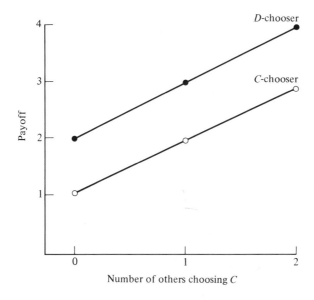

FIG. 8.1   Graphical representation of uniform NPD.

The game can be represented in graphical form without loss of information (see Figure 8.1). Note that the horizontal axis indicates the number of *other* players choosing C and the payoff functions are those of the remaining player depending on whether he chooses C or D. This enables a comparison to be made between an individual player's payoff if he chooses C and his payoff if he chooses D for any specified number of C choices on the part of the other players. The fact that the payoff functions are parallel indicates that the D strategy dominates the C strategy by a constant amount in this case: a D choice yields exactly one unit of payoff more than a C choice irrespective of the number of others choosing C. The left-hand extremity of the D-chooser's function is clearly the only equilibrium point since a player will always have cause to regret a C choice. But the right-hand extremity of the C-chooser's function, which represents the payoff to a C-chooser when everyone else chooses C, is higher than the equilibrium point, indicating that the equilibrium is deficient. The example in Figure 8.1 is the simplest possible NPD structure, and it can easily be extended for any (finite) number of players: the overall form is preserved but the payoff functions are lengthened. For NPDs with non-parallel and non-linear (curved) payoff functions, see Schelling (1973).

Matrix 8.3

		II	
		C	D
I	C	3/2, 3/2	1/2, 2
	D	2, 1/2	1, 1

The formal relationship between the NPD and the PDG is evident from a careful examination of Matrix 8.3. The game represented in Matrix 8.3 is a conventional two-person PDG. Suppose that each of three players plays this game with each of the other two players. The resultant *compound game* then turns out to be the NPD depicted in Matrix 8.2 and Figure 8.1. If all three players choose C, for example, each receives a payoff of 3/2 in the game against one of his opponents and 3/2 in the game against the other; each player's total payoff is therefore 3/2 + 3/2 = 3 units, which agrees with the corresponding NPD outcome. If two players choose C, then each C-chooser gets 3/2 against the other C-chooser and 1/2 against the D-chooser, making a total of 2 units, which also agrees with the NPD outcome, and so on. The three-person game may be thought of as a compound game based upon the two-person PDG shown in Matrix 8.3. A compound game involving any number of players can be constructed on the basis of the PDG, and it always turns out to be an NPD with the overall form shown in Figure 8.1. If the PDG is non-decomposable, however (see Section 7.4), the payoff functions of the corresponding NPD will not be parallel.

A simple algebraic model can be used to represent a decomposable Prisoner's Dilemma (PDG or NPD). Let each of the N players ($N = 2$ in the PDG) receive an amount c for choosing C and an amount d for choosing D. In addition, each player is fined an amount e for every player who chooses D.

In the game shown in Matrix 8.2 and Figure 8.1, $c = 3$, $d = 5$, and $e = 1$. In the two-person case shown in Matrix 8.3, $c = 3/2$, $d = 3$, and $e = 1$.

The defining properties of the generalized decomposable Prisoner's Dilemma are the following inequalities:

$$d - e > c > d - Ne > c - (N - 1)e,$$

which are simply the familiar inequalities $T > R > P > S$ used to define the PDG in Section 6.7. Simplifying these inequalities, we find that $N > 1$—the number of players must be two or more—and combining them, we arrive at

$$Ne > d - c > e,$$

which may be regarded as a generating formula for decomposable Prisoner's Dilemmas of any size.

Comparing the NPD with the conventional PDG, Davis, Laughlin, and Komorita (1976) commented:

> The N-person case (NPD) has greater generality and applicability to real-life situations. In addition to the problems of energy conservation, ecology, and overpopulation, many other real-life problems can be represented by the NPD paradigm. . . . It seems reasonably safe to predict that we will see an increasing number of studies based on the NPD (pp. 520–1).

The experimental literature in this area is reviewed in Chapter 9.

8.8   General Theory of Compound Games

The NPD is a special case of a compound game based upon a 2×2 (two-person, two-choice) game. In this section we shall briefly examine some other compound games and derive some general results. The mathematical theory of compound games does not seem to have been developed in detail by previous writers. It will be assumed that the multi-person game and its underlying two-person game are symmetric in the sense of being the same from every player's viewpoint, and that the payoff resulting from a C or a D choice is a linear function of the number of other players choosing C.

Matrix 8.4

		II	
		C	D
I	C	R,R	S,T
	D	T,S	P,P

The generalized form of a symmetric 2×2 game, introduced in Section 6.7, is reproduced in Matrix 8.4. The players in the multi-person game each play a 2×2 game of this type with each of the others. Considering the multi-person game from a single player's viewpoint, let the number of *other* players be denoted by N (strictly speaking, we are dealing with an $N + 1$-person game). Let the number of other players choosing C be x. The total payoff to a player choosing C, denoted by $P(C)$, and the total payoff to a player choosing D, denoted by $P(D)$, are then defined by the following payoff functions:

$$P(C) = Rx + S(N - x),$$
$$P(D) = Tx + P(N - x).$$

The values of the $P(C)$ and $P(D)$ functions at their end-points are found by setting $x = 0$ and $x = N$. Thus if none of the other players chooses C, that is, if $x = 0$, the payoff to a solitary C-chooser is SN and the payoff to a D-chooser is PN. If all of the other players choose C, that is, if $x = N$, a C-chooser gets RN and a solitary D-chooser is TN. It is clear that, in the case of the NPD, TN can be interpreted as the temptation to be the sole D-chooser, RN the reward for joint C-choices, PN the punishment for joint D-choices, and SN the sucker's payoff for being the sole C-chooser.

Four multi-person compound games are depicted graphically in Figure 8.2. Figure 8.2a shows a multi-person game based upon a 2×2 matrix with $T > S > R > P$, referred to as Leader in Section 6.7; 8.2b is a Multi-Person Battle of the Sexes with $S > T > R > P$; 8.2c is a Multi-Person Chicken Game with $T > R > S > P$; and 8.2d is a modified Multi-Person Maximizing Difference game with $R > T > P > S$. Equilibrium points are indicated by dashed circles.

In cases (a), (b), and (c), the equilibrium points are at the intersections of the payoff functions. To see why this is so, consider first the points to the left of the intersection in one of these graphs; this shows the choice facing a player when few of the others choose C (x is small). In each case the individual player would regret a D choice and would switch to a C choice if the game were repeated, since the C function is above the D function in this region. Thus D-choosers will tend to become C-choosers and the outcome will move to the right as x increases. To the right of the intersection, exactly the reverse holds: C-choosers will switch to D and the outcome will move to the left as x decreases. At the intersection, and only there, no player will have cause to regret his choice and none will be motivated to switch since the payoff is not improved by switching unilaterally, given the indicated number of C-choosers (the value of x at this point). The intersection is therefore an equilibrium point, and any deviation from it will tend to be self-correcting.

In each of the games shown in Figure 8.2(a), (b), or (c), the payoff at the equilibrium point is the same whether the individual player chooses C or D.

The two payoff functions can therefore be equated at this point:

$$Rx + S(N - x) = Tx + P(N - x).$$

In 101-person Chicken (where the number of "other" players is $N = 100$), for example, suppose $T = 4$, $R = 3$, $S = 2$, and $P = 1$. Then:

$$3x + 2(100 - x) = 4x + (100 - x),$$
$$x = 50,$$

and substituting this value of x in either payoff function, the payoff turns out to be 250. With these parameters, therefore, if the game is repeated, there

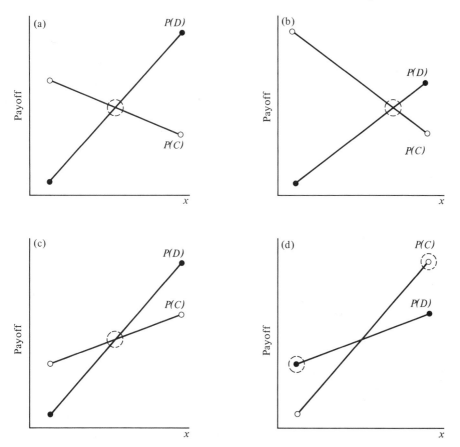

FIG. 8.2   Multi-person compound games based on 2 × 2 matrices. Panel (a) is Multi-Person Leader, (b) is Multi-Person Battle of the Sexes, (c) is Multi-Person Chicken, and (d) is a modified Multi-Person Maximizing Difference game. The $P(C)$ functions indicate the payoffs to a player choosing C when x of the other players choose C. The $P(D)$ functions indicate the payoffs to a player choosing D when x of the others choose C. Dashed circles indicate stable equilibrium points.

will be a tendency towards a stable equilibrium with 50 players choosing *C* and all players receiving payoffs of 250 units. Societies will therefore tend to evolve towards a state of affairs in which some fixed proportion of individuals adopt threatening types of behaviour towards others in Chicken-type encounters. If all such individuals are locked away, others are bound to take their places. As in the NPD, the equilibrium is deficient since everyone would be better off if all chose cooperative *C* strategies, but this outcome is unstable. Multi-Person Chicken is, however, unlike the NPD in so far as the *D* function does not dominate the *C* function across its entire length.

In Figure 8.2(d) we have a different case. This is a modified Multi-Person Maximizing Difference Game called "Backpatting" by Hamburger (1979, pp. 167–8) and it has two stable equilibrium points. It is better to resist than not to resist an invading army if everyone else resists, but it is better not to resist if no-one else resists, and the first of these "scenarios" is better for everyone than the second. If everyone chooses *D* or if everyone chooses *C*, no-one is motivated to switch. If only a few choose *C*, that is, when *x* is small, they will switch to *D*, and if most choose *C* the few who choose *D* will switch to *C*: the tendency is always *away* from the intersection point. In a society in which most people are competitive in encounters with this structure, it is better to compete than to cooperate, but it is better to cooperate if most other people are cooperative. But the equilibria are not equivalent: everyone is better off in a society in which all choose *C* than in one in which all choose *D*. The direction in which a society will evolve depends on the initial proportion of *C*-choosers; an initial bias in one direction or the other will tend to be self-reinforcing.

8.9   Summary

The distinction between cooperative multi-person games in which binding agreements and coalition formation among the players are permitted, and non-cooperative games in which they are not, was explained in Sectioin 8.1. Section 8.2 was devoted to non-cooperative theory, which is based on equilibrium points, and the "solutions" were shown to be weak in many cases. Section 8.3 centred on cooperative theory, which makes use of a mathematical formalization, known as the characteristic function, of the value of coalitions to their members. In Section 8.4 Harold Pinter's play, *The Caretaker*, was used to illustrate the shifting pattern of coalitions in a three-person game with an empty core; that is, one in which a potential coalition exists with the will and the power to replace any existing coalition. The Shapley value, a measure of the players' *a priori* power in a game, was introduced in Section 8.5, and interesting and unexpected results were shown to result from its application to the United Nations Security Council and other real decision making bodies.

Section 8.6 was concerned with the Dollar Auction game, a frustrating multi-person game which models the dynamic features of escalation and the Concorde fallacy (the tendency to continue investing in a project simply to justify past investment). The N-Person Prisoner's Dilemma, "discovered" in the mid-1970s, was discussed at length in Section 8.7. This game models problems of energy and resource conservation, inflation and voluntary wage restraint, environmental pollution, overpopulation, multilateral disarmament, and many others besides. Its most important property is that players who all pursue their individual self-interests end up worse off than if they had all acted against their individual self-interests. In Section 8.8 a general theory of compound games was sketched. It was shown that societies have a tendency to evolve towards stable states in which all, none, or some fixed proportion of individuals act cooperatively towards each other in two-person interactions with specified strategic structures.

9

Experiments With Coalition, Auction, and Social Dilemma Games

9.1 Multi-Person Experimental Games

THROUGHOUT the 1960s and early 1970s experimental gaming was devoted overwhelmingly to two-person games while multi-person games were largely neglected. One of the reasons for this imbalance was that, before the mid-1970s, no multi-person games of comparable psychological interest to the two-person Prisoner's Dilemma had been identified by game theorists. During this period the only multi-person experimental gaming that took place was devoted to coalition formation in groups in which communication, bargaining, and negotiation were permitted, but this line of investigation did not attract a large following.

The Dollar Auction game (Shubik, 1971), in spite of its obvious psychological significance, did not immediately stimulate many empirical investigations on account of the formidable methodological problems of converting it into a usable laboratory game; but ways were eventually found round these problems and some interesting research has begun to emerge in this area. The Multi-Person or N-Person Prisoner's Dilemma (Schelling, 1973; Hamburger, 1973) and the strategically equivalent Commons Dilemma game (Dawes, 1973), on the other hand, are not only psychologically interesting and applicable to a wide range of socially significant strategic interactions, but also comparatively easy to convert into simple laboratory games. Empirical investigations of social dilemmas consequently developed into an active area of research almost immediately after the early theoretical analyses were published.

In the following section, reseach on coalition formation in cooperative multi-person games is selectively reviewed. Section 9.3 contains a summary of all published experiments on auction games and related "psychological traps". In Section 9.4 the literature on N-Person Prisoner's Dilemma experiments is examined, and in Section 9.5 a new experiment comparing behaviour in abstract and structurally equivalent lifelike versions of this game is reported. Section 9.6 contains a brief summary of the chapter.

9.2   Coalition Formation

This is an area that has become rather top-heavy with theory: a large number—literally dozens—of theories of coalition formation have been proposed but the supporting evidence is relatively slender and, until quite recently, has been based largely on studies of triads (three-person groups). Critical tests of competing theories, however, often call for experiments with groups larger than the triad (Komorita and Chertkoff, 1973). The literature will not be exhaustively reviewed in this section since a number of comprehensive summaries are available elsewhere; among the most useful are those by Gamson (1964), Vinacke (1969), Chertkoff (1970), Stryker (1972), Tedeschi, Schlenker, and Bonoma (1973, chap. 6), and Burhans (1973). More recent evidence has been commented upon briefly by Davis, Laughlin, and Komorita (1976, pp. 523–4) and by Hamburger (1979, pp. 243–5).

The most popular research methodology, pioneered by Vinacke and Arkoff (1957), makes use of a *pachisi* (or *parcheesi*) board. Pachisi is an Indian board game similar to backgammon, ludo, or monopoly. In coalition experiments, each member of a triad (or larger group) places a counter on the START square of the board and moves it a certain number of squares towards HOME each time a die is cast. Relative weights—analogous to voting strengths in a shareholder's meeting or a political assembly, for example—are fixed by assigning a specific number to each player. If, let us say, the three players are weighted 4, 3, and 2 respectively and the die shows 6, the first player moves 24 (4 × 6) squares towards HOME while the others move 18 (3 × 6) and 12 (2 × 6) squares respectively. The object of the game is to get to HOME before the other players.

The rules of the game allow communication and bargaining so that a pair or a group of players can negotiate a coalition at any point. A coalition receives a new counter and moves as a single player with a weight equal to the sum of its constituent members' weights; if the coalition members are weighted 4 and 3, for example, then their combined weight is simply 7. A pair or a group wishing to form a coalition has to commit itself in advance to an agreed division of the payoff among its members should the coalition win. The first player or coalition to reach HOME receives a payoff (in points or money) and the others receive nothing.

Other experimental procedures apart from the pachisi board game are possible. Gamson (1961a), for example, was the first to use a simulated *convention* game in which subjects play the roles of delegates, each controlling a fixed number of votes, in the manner of bloc delegations at political party conventions in the United States or representatives at the Council of Ministers of the European Economic Community.

Most experiments have centred upon the effects of different distributions of weights on the emergence of particular coalitions predicted by different theories, and on the divisions of payoffs agreed among the members of coalitions. Some of the more prominent theories and findings are discussed below.

Game theory

From the characteristic function of a cooperative game (see Section 8.3) it is possible to see at a glance which of the coalitions are winning combinations and which are not, provided that it is a simple game—that is to say, a win–lose game. If, for example, there are three players, I, II, and III weighted 4, 3, and 2 respectively, it is clear that any two-person coalition or the grand coalition of all three players can guarantee to win. A majority of the resources (voting strengths or weights) is controlled by any two-person or three-person coalition, whereas a single player acting alone can be defeated by a coalition of the other two. If the weights are 4, 2, and 1, on the other hand, then (I), (I, II), (I, III), and (I, II, III) are the winning coalitions since each has a majority; the (II, III) coalition has insufficient resources to win.

According to game theory, one of the winning coalitions ought to form if the players are rational and if they play solely according to the prescribed payoffs. But game theory is normative rather than positive: it makes no predictions about how ordinary human subjects *will* play, and it is therefore incapable of being refuted by experimental evidence. It is a common mistake to interpret experimental findings as showing that "the implication of rational [game theory] that every sensible person ought to act in a formally prescribed fashion, given certain stated conditions, is unjustified" (Vinacke, 1969, p. 313), since the justification is logical rather than empirical. In any event, the overwhelming weight of evidence (e.g. Vinacke and Arkoff, 1957; Kelley and Arrowood, 1960; Stryker and Psathas, 1960; Psathas and Stryker, 1965; Chertkoff, 1966, 1971) suggests that winning coalitions, defined according to the characteristic function, form much more often than losing coalitions. This is consistent with the view that subjects generally act rationally according to this game theory criterion.

Minimal winning coalition theory

William Riker (1962, pp. 40–6, 247–78) argued that, given certain reasonable assumptions, no incentives exist for coalitions larger than the minimal winning size to form. The ejection of superfluous members allows the payoff to be divided among fewer players, and this is bound to be to the advantage of the remaining coalition members (see also Riker, 1967, pp. 167–74; Riker and Ordeshook, 1973, pp. 179–80). This theory is often referred to as the *size principle*. If the players' weights are 4, 3, and 2 in a triad, the theory predicts that (any) two-way coalition will form, while in the 4, 2, 1 case no successful bargains will be struck since Player I can win the whole payoff alone and has no incentive to take on any partner. The theory is clearly more specific than game theory in so far as it singles out one or more of the winning coalitions defined by game theory as the likely outcome, but it is not as powerful in this respect as certain other theories.

Stringent tests of minimal winning coalition theory require the use of groups containing four or more players, and the evidence from such experiments tends to corroborate the theory strongly. One of the most ambitious and carefully controlled studies was conducted by Michener *et al.* (1975), who reported that coalitions without superfluous members formed in 88 per cent of cases out of a total of 288 plays with various weight distributions in four-person groups. The grand coalition of all players accounted for most of the remaining outcomes reflecting, perhaps, anti-competitive attitudes among some subjects.

Riker (1962) and Brams (1975, pp. 220–32) have summarized empirical evidence from the field of national and international politics bearing on minimal winning coalition theory. In the international arena the historical record shows that overwhelming—and hence obviously non-minimal—coalitions, such as the Allied powers after World Wars I and II, were invariably plagued by internal strife. According to Brams (1975), "the hopes for permanent peace enshrined first in the League of Nations and now in the United Nations founder on the size principle precisely because of the undiscriminating inclusiveness of these international organizations" (p. 223). Further anecdotal evidence comes from the disintegration of superfluous majorities in American politics, and examples have more recently been adduced from French, Danish, Italian, Dutch, West German, Israeli, Japanese, and even tribal African politics (for references, see Brams, 1975, p. 227). Most of this evidence seems to confirm Riker's minimal winning coalition theory.

Minimum resource theory

This theory, proposed by William Gamson (1961b, 1964), does not contradict the predictions of minimal winning coalition theory, but it is more specific in certain cases and therefore more powerful. Consider once again the case of a triad in which Players I, II, and III are weighted 4, 3, and 2 respectively. Game theory establishes that in this case *any* two-person or three-person coalition is rational; minimal winning coalition theory predicts that one of the *smallest* winning coalitions (I, II), (I, III), or (II, III) will form; while minimum resource theory singles out the *cheapest* winning coalition (II, III) as the likely outcome. This last-mentioned coalition is the one "in which the total resources are as small as possible while still being sufficient" and it arises, according to the theory, from a "parity norm", that is, a belief likely to be shared by the players "that a person ought to get from an agreement an amount proportional to what he brings into it" (Gamson, 1964, pp. 86, 88).

If the players are guided by the parity norm, they are inevitably led to the winning coalition that controls the minimum resources. In the 4, 3, 2 case, Player I will prefer to join with III rather than with II because he can expect a larger share of the payoff from the coalition in which his weight is relatively larger in comparison with his partner's. For the same reason, II will prefer to join with III, and III will prefer to join with II. The only one of these choices of partner that is reciprocated is the one between Players II and III who will therefore, according to the theory, form what (inevitably) turns out to be the winning coalition that controls the minimum resources.

Experimental investigations (e.g. Vinacke and Arkoff, 1957; Chaney and Vinacke, 1960; Kelley and Arrowood, 1960; Chertkoff, 1966, 1971; Crosbie and Kullberg, 1973; Chertkoff and Braden, 1974) have tended to confirm the predictions of minimum resource theory about which coalitions are most likely to form. These results are particularly impressive in view of the fact that they are often counterintuitive: in the example above, the player who seems to be most powerful (Player I) tends to be excluded from coalitions precisely because of his "excessive" power and is actually at a disadvantage compared to the weaker players.

The exclusion of the apparently strongest player from coalitions in games with certain weight distributions is called the *power inversion* paradox. It makes certain otherwise obscure political and economic phenomena intelligible, and provides a strategic interpretation of the way in which people tend to gang up against one who seems to be too powerful. But it is important to realize that the power inversion paradox arises from an illusion; it has no rational basis. In the 4, 3, 2 example, the players in reality have equal power in spite of their disparate weights: *any* two-person coalition wins and the weights are only window-dressing; a Shapley value analysis (see Section 8.5) confirms that the players' *a priori* power is equal. Several researchers have found that the power inversion effect declines when a group of subjects play a simple coalition game of this type many times over; the subjects gradually come to realize that the weights do not reflect real power differences (Kelly and Arrowood, 1960; Stryker and Psathas, 1960; Psathas and Stryker, 1965).

Minimum resource theory provides excellent predictions about coalition formation, but the theory's predictions about the distributions of payoffs to which coalition members will agree are not well supported by experimental evidence. According to the theory, the players should split the payoffs by applying the parity norm, that is, in proportion to their relative weights. In the 4, 3, 2 example, if Players II and III form a coalition—as they usually do—II should get 3/5 and III 2/5 of the payoff by mutual agreement. But the evidence consistently shows that subjects tend to reach agreements somewhere between this parity split and an equal division of the payoff (e.g. Kelley and Arrowood, 1960; Chertkoff, 1971). Some researchers think that these findings refute minimum resource theory (e.g. Wilke and Mulder, 1971) while others consider them to be consistent with the theory (e.g. Crosbie and Kullberg, 1973).

Gamson (1962) tested his minimum resource theory on historical data from real Presidential nominating conventions over a 53-year period. These data support the theory reasonably well but the fit is not perfect. Gamson commented that the "modest success here in situations characterized by so many unique historical features is more satisfying than the neater but more easily obtained experimental results" (p. 171).

Numerous other theories of coalition formation and payoff apportionment have been proposed, including an *utter confusion* theory (Gamson, 1964) which predicts that accidental and irrelevant factors like loudness of voice and seating pattern will determine coalition choices in difficult circumstances. Random coalitions are predicted when time pressure is severe, there are numerous players, the players are strangers and are unsophisticated, the rules of the game are complicated, communication difficulties exist, the payoff is barely worth striving for, and so forth. The evidence tends to confirm the obvious prediction of utter confusion theory that coalitions will form more or less at random in such circumstances (e.g. Kalisch *et al.*, 1954; Willis, 1962). At the opposite end of the spectrum, when conditions are favourable and the players are strongly motivated, game theory, minimal winning coalition theory, and minimum resource theory all generate excellent predictions about coalition formation. The ways in which coalition members distribute payoffs remain somewhat problematical, however; notions of equity appear to operate in complex ways when it comes to splitting the payoffs.

9.3 Auction Games and Psychological Traps

The use of auction games (described in Section 8.6) as laboratory tasks presents a number of methodological problems. If real money is used, then the payoff structure of the game is likely to be ambiguous because the subjects may not believe that they are liable to be fleeced by the experimenter. The experimenter could, of course, take pains to convince the subjects that they stand to lose money, but an experiment that produces a financial profit at the expense of volunteer subjects seems ethically dubious. If, on the other hand, the subjects play for points or valueless tokens rather than real money, then an essential psychological ingredient of the game is missing: subjects are less likely to continue bidding merely in order to justify past investments if these investments are trivial or meaningless to them.

Allan Teger (1980) and his colleagues got round these problems in a number of ways. In their informal field experiments, conducted in the form of classroom demonstrations with post-experimental interviews, the subjects were invited to bid with their own money and were told that credit would be extended to anyone who had insufficient ready cash to pay outstanding debts at the end of the auctions. The experimenters did not, in fact, accept any payments from the subjects after the auctions, although the post-experimental interviews showed that the subjects believed that they would have to pay up.

In Teger's (1980) experiments with signed-up volunteers it was necessary to give each subject an initial monetary stake with which to bid. Subjects normally expect to be paid for participating in experiments and they would probably not have believed that they could possibly lose their own money. The technique of providing subjects with initial stakes with which to bid was pioneered by Tropper (1972), who was the first to report an experiment using an auction game (it is surprising that Teger and his colleagues have not cited Tropper's work). It is not an entirely happy solution because the money that the subjects stand to lose is, in a sense, not really their own: they did not have it before entering the laboratory, and losing it may not have the same psychological significance as losing money that is unambiguously theirs. The amount given to the subjects turned out, in fact, to be a powerful determinant of bidding behaviour in Teger's experiments.

A third approach adopted by Teger (1980) and his colleagues in some of their laboratory experiments was this: each subject was given 975 points with which to bid for a prize of 500 points, and was told that each point was worth an unspecified amount of money to be cashed at the end of the auction. The subjects did not know that their bidding partners had also been assigned an initial stake of 975 points. The idea behind this arrangement was "that the subjects not feel that the money we gave them was inconsequential (and thus that they had nothing to lose), or that it was a considerable sum (and thus that they had best quit immediately and take the money without bidding)" (p. 20). The argument is not entirely clear, but the procedure worked in practice.

The results of auction game experiments are surprising and edifying. Shubik (1971) was the first to report that the game "is usually highly profitable to its promoter" (p. 109). Tropper (1972) conducted the first controlled experiment in this area, in which the items auctioned were a dollar bill, a felt-tipped pen worth 49 cents, and a pocket flashlight worth 99 cents. Each subject was given 250 cents with which to bid, and in 16 out of 30 auctions, apparently conducted with pairs of subjects, one or both of the bidders paid more than the (subjectively rated) value of the prize.

In approximately 40 informal field experiments with groups of undergraduates, graduates, and faculty members as subjects, Teger (1980, chap.

2) found that the bidding for a dollar bill *always* exceeded 1 dollar and sometimes went as high as 20 dollars. Subjects who were caught in the spiral of escalation often became extremely emotional, crying, sweating, and glancing anxiously around for reassurance. One student began bidding to rescue his girlfriend who had bid over a dollar and had told him that she was scared: "He entered the bidding, figuring that their combined resources would enable them to outlast the other bidder" (p. 16). Many subjects interpreted what had happened in a remarkably egocentric way, describing their opponents in post-experimental interviews as "crazy" for bidding more than the prize was worth in spite of having done so themselves. Some subjects made bids in spite of having seen the auction being conducted in previous experiments! They claimed that it had not occurred to them that they could be caught in the same trap as the others.

The laboratory experiments reported by Teger (1980, chaps 2, 3, 4, and 5) and his colleagues were conducted with pairs of bidders rather than larger groups. Even under these conditions the bidding exceeded the value of the prize in a substantial proportion (40 to 50 per cent) of cases, which replicates Tropper's (1972) findings mentioned earlier. Perhaps the most striking discovery was that in those pairs in which the 1-dollar threshold was crossed the bidding usually continued to escalate until the subjects' resources were *entirely* depleted. This suggests that the value of the prize serves as a psychological "point of no return" for most bidders, beyond which they are unable or unwilling to extricate themselves from the game until they are forced to do so. This phenomenon was observed time and time again, whether the subjects were playing with real money or points to be converted into money at an unknown exchange rate after the auction. It is sufficiently striking to deserve a dramatic label; perhaps the *Macbeth effect* is appropriate ("I am in blood / Stepped in so far that, should I wade no more, / Returning were as tedious as go o'er"—*Macbeth*, Act III, Scene iv).

Subjects given the opportunity to take part in two successive laboratory auctions showed remarkably little evidence of learning from experience: the distributions of bids were similar to those of inexperienced pairs of subjects and the Macbeth effect was not suppressed. This finding is important because it shows that the phenomenon of escalation is not due merely to a failure on the part of the subjects to appreciate the danger inherent in the game or the possibility that the bidding might exceed the value of the prize. More interesting psychological processes are evidently at work.

Teger and Carey (1980) investigated the thought processes of subjects during auction games and discovered that an important motivational change takes place in most cases. When making a bid at any stage in the experiment the subjects were requested to indicate on a set of seven-point rating scales the importance of various motives underlying their decisions to bid: "in order to make money", "in order to show that I am better at the task than the other person", "in order to regain some of the money that I have lost", and so on. The results showed that

> the initial bidding appears to be motivated by economic concerns, but the tendency to bid until you are broke is due to a new motivation which develops during the course of the auction—a motivation toward competition which makes the ecnomic considerations less important (p. 60).

The (interpersonal) competitive motive, which appears to stem from face-saving considerations, tends to supplant the (purely individual) economic motive when the bidding exceeds the value of the prize; it is at this point that interpersonal considerations tend to assume major significance.

Teger *et al.* (1980) monitored a number of physiological stress indicators in subjects taking part in auction games. The results of this experiment confirmed the finding of previous investigations that the most significant "moment of truth" for the subjects occurred when the bidding first passed the value of the prize. In particular, a sharp decrease in heart rate was usually recorded at this point. The physiological pattern observed in subjects at this "point of no return" is similar to the pattern observed in parachutists just before jumping from an aircraft. It may be due, at least in part, to the subjects' sudden realization that it is possible for the bidding to continue beyond the 1-dollar point; a controlled experiment with experienced or pre-educated subjects would be necessary to discover to what extent this factor is important.

Since the auction game is, in a sense, a product of the Vietnam war, it is not unreasonable to ask what light it throws on that devastating example of escalation. A content analysis of the speeches of President Johnson on Vietnam from 1964 to 1968 (Sullivan and Thomas, 1972) invites some very tentative conclusions. In the early stages of military escalation, the symbolic words used by Johnson were reportedly mainly "positive" (*democracy, freedom, liberty, justice*), but later phases of escalation were associated with the use of "status" words (*honour, will, status*). It is conceivable that this change may reflect a motivational change in Johnson's approach to the conflict, and it may parallel the motivational change in subjects in auction game experiments, from gaining financially to saving face by beating the other bidder even if it entails enormous loss to both. The "moment of truth" in the Vietnam war may have been reached when the original "positive" goal of a "clean" victory was abandoned in favour of saving face in spite of the massive costs involved. There is a danger of psychologism in pursuing this analogy too far, however.

Carey (1980) has argued that auction games are applicable to many naturally occurring strategic interactions, and he has suggested a number of examples based on industrial relations, divorces, and fights. The Macbeth effect, for example, is clearly discernible in the records of industrial strikes at some locations. Carey's claim that "nearly every interpersonal situation can be seen to have the basic features of the dollar auction game" (p. 129) is an obvious example of theoretical imperialism, but the relevance of the game to certain political, economic, military, and interpersonal interactions is striking. The most horrifying example of all is surely the escalating arms race between the nuclear powers, which seems to have all the formal properties of an auction game (Rapoport, 1971).

Jeffrey Rubin and Joel Brockner have reported a series of experiments on *psychological traps* that are closely related to auction games. In these experiments subjects were invited to invest escalating amounts of time and money in the hope of obtaining valuable prizes. They were free to cut their losses and withdraw at any point, but by doing so they had to forgo the possible prizes. Situations of this sort arise naturally when, for example, one has spent a long time waiting for a bus; the longer one has waited, the less one is inclined to leave the bus stop. This phenomenon is a close cousin to the Concorde fallacy described in Section 8.6, and can be explained by cognitive dissonance theory (Festinger, 1957). There is an uncomfortable inconsistency between the knowledge that one has invested heavily in something and the belief that it is not worth so much time, effort, or money. One way of reducing the psychological inconsistency—and hence the discomfort—is by developing a more favourable attitude towards the prize. As the investment escalates, one's subjective evaluation of the prize may become unrealistically inflated.

The first experiment in the series (Rubin and Brockner, 1975) provided strong evidence of the entrapment phenomenon. The subjects were each given an amount of money before the experiment began and were promised an additional (much larger) amount if they succeeded in solving a crossword puzzle. But the subjects had to invest money in order to try for the prize: after working on the puzzle for a few minutes they were fined 25 cents for every subsequent minute that elapsed. They were permitted to abandon the puzzle at any point and to retire with what was left of their initial payments. Those who failed to do so found that their initial payments were soon depleted; they were effectively paying out of their own pockets for remaining in the experiment. The puzzle was too difficult to solve without a dictionary, which the subjects were told was available on a "first come, first served" basis, but circumstances were arranged so that no subject ever got to the front of the queue. While queueing for the dictionary they were not allowed to continue working on the puzzle.

Approximately 20 per cent of Rubin and Brockner's (1975) subjects remained in the queue beyond the point at which their initial payments were depleted and they went into debt. They waited longer, on average, when the fines mounted rapidly rather than slowly, when they were not required to keep careful records of how much money they had lost, and when they thought they were near the front of the queue. The experiment provides a vivid illustration of psychological entrapment. It brings to mind W. C. Fields's famous motto: "If at first you don't succeed, try again. Then quit. No use being a damn fool about it" (quoted in Cohen and Cohen, 1980, p. 114). Subsequent experiments confirmed and extended the earlier findings. Brockner, Shaw, and Rubin (1979) investigated limit-setting and active versus passive decisions to continue investing in an entrapment game. Some of the subjects were required to inform the experimenter in advance of their provisional (non-binding) limits, indicating how much they were prepared to invest; another group had to set private limits; and a third group were not required to set any limits. In each group, half of the subjects had to make repeated active decisions to remain in the experiment while the others remained unless they made active decisions to quit. This latter condition is reminiscent of inertia selling, in which a customer is deemed to have agreed to buy a product unless some positive action is taken. The results showed that subjects became most severely entrapped when they were not required to set prior limits on investment and when they had to make active decisions to quit.

Rubin *et al.* (1980) reported further evidence indicating that entrapment tends to be greater under passive inertia-selling conditions than when active decisions have to be made to continue investing, although this effect failed to emerge from one of the two experiments reported in this paper. The other experiment yielded evidence suggesting that male subjects are more vulnerable to entrapment than females.

Brockner, Rubin, and Lang (1981) reported two further experiments linking entrapment with face-saving. The results of both experiments revealed that "individuals will become more or less entrapped to the extent that doing so will portray them in a favorable light" (p. 78). The finding is consistent with the results of experimental auction games mentioned earlier. The experiments also revealed that entrapment tends to be reduced if the subjects' attention is directed to the costs of investing rather than the value of the prize. This was achieved by reminding some of the subjects of the advantages of saving money and requiring them to keep charts of their investments continuously up to date.

9.4 *N*-Person Prisoner's Dilemma

The essential properties of the *N*-Person Prisoner's Dilemma (NPD), discussed at length in Section 8.7, are as follows: (a) each player faces a choice between two pure strategies, which may be labelled *C* (cooperate) and *D* (defect); (b) the *D* strategy is dominant for each player, so that each is better off choosing D than C irrespective of how many of the other players choose C; and (c) the outcome is preferable from every player's point of view if all choose C than if all choose D, in other words the dominant D strategies intersect in a deficient equilibrium.

The NPD became established as an object of active empirical investigation in the mid-1970s, and the first major review of experimental evidence was provided by Dawes (1980). The review that follows in this section is intended to complement rather than to duplicate Dawes's; it is not exhaustive but it includes some important studies omitted from the earlier review.

A classic experiment by Mintz (1951) on "non-adaptive group behaviour", which was designed as a laboratory analogue of escape panics, is the earliest precursor of modern NPD research. It has not been mentioned in previous discussions of the NPD but it is clearly pertinent since the strategic structure of Mintz's task corresponds to the NPD, given certain reasonable assumptions. The basic methodology was simple and ingenious. Groups of subjects were given the task of removing aluminium cones from a narrow-necked bottle as quickly as possible; each subject held one end of a string, the other end of which was attached to one of the cones. In some treatment conditions the bottle was slowly filled with water from below and the subjects were each offered a substantial monetary reward for extracting their cones before they became wet. There was enough time for all subjects to remove their cones in an orderly fashion, but if two or more cones arrived at the neck of the bottle together a blockage was bound to occur.

In the conditions involving water and monetary incentives, the task is strategically equivalent to the situation that occurs, for example, in a crowded theatre when a fire breaks out and members of the audience are strongly motivated to escape quickly via a narrow exit. Mintz's (1951) water represents the fire, the neck of the bottle corresponds to the theatre exit, and the cones are the members of the audience. It is not unreasonable to make the simplifying assumption that each individual faces a choice between filing out in an orderly fashion and rushing for the exit. Under this assumption the strategic structure of a Mintz-type escape panic is an NPD because from an individual point of view one's chances of survival are better—irrespective of the choices of the others—if one rushes (D), yet everyone's chances are better if everyone files out in an orderly fashion (C).

The invariable outcome in Mintz's (1951) experiment was a traffic jam at the neck of the bottle, preventing all or some of the cones from being removed. When subjects were allowed to discuss the task and plan their behaviour in advance there was some improvement; however traffic jams still usually resulted. But when the water and the financial incentives were removed and the subjects were instructed merely to withdraw their cones as quickly as possible, serious blockages never occurred. The results are perfectly in accordance with the theory of the NPD. In the water and incentive conditions the rushing strategy is dominant and remains so even if (as a result of prior communication) the others' intentions are known in advance. Removal of the water and the incentives, on the other hand, amounts to altering the strategic structure of the underlying game, since rushing for the exit is no longer the dominant strategy; the task becomes a pure coordination game in which the only problem is to agree upon an order in which to file out, and there are ample opportunities for tacit communication to achieve this end (first come, first served is a culturally prominent solution).

More recent experiments have made use of explicit NPD structures in which the subjects' available options are unambiguously identified and the payoff functions are presented numerically, usually in the form of tables of outcomes and corresponding payoffs. The major dependent variable has usually been the proportion of cooperative (C) choices, and attention has focused on the effects of group size, communication between players, and attributions of intent. Other independent variables have generated less interesting results; the sex differences found in two-person PDGs, for example, does not appear in the NPD (Caldwell, 1976; Goehring and Kahan, 1976; Dawes, McTavish, and Shaklee, 1977).

Studies of the effects of group size on cooperation in the NPD present methodological problems which have not been satisfactorily solved by all investigators. The problems arise from the fact that it is impossible to hold all aspects of an NPD constant while varying the number of players. If the payoff to a C-chooser when all other players choose C, the payoff to a D-chooser when all other players choose D, and the individual advantage of choosing D rather than C are equated in groups of different sizes, for example, then the extent to which a C choice benefits each of the other players will not be the same.

Different parameters have been held constant by different investigators and, with the exception of one of the three games used by Bonacich et al. (1976), less cooperation has always been found in larger than smaller groups. Marwell and Schmitt (1972), for example, compared two-person and three-person groups: "The basic data clearly support the hypothesis that rates of cooperation are inversely related to the number of people involved in the interaction" (p. 379). Hamburger, Guyer, and Fox (1975) reported similar results in a comparison of three-person and seven-person groups, and Fox and Guyer (1977) found substantially more cooperation in three-person than in seven-person or 12-person groups. Hamburger (1977) attempted to control all factors apart from group size by introducing probabilistic payoffs and using stooges to fill out the larger groups: pairs of genuine subjects played either a two-person PDG or what appeared to be a three-person NPD including the third stooge player. The stooge chose in such a manner that the parameters of the two-person and three-person games were identical from the point of view of the genuine subjects. Despite the objective equivalence of the two games, the subjects who thought they were playing the three-person game cooperated significantly less frequently than those in the two-person game.

How can the lower levels of cooperation in larger groups be accounted for? One of the most plausible explanations is the *bad apple* theory. This theory rests on the assumption that it takes only a few *D*-choosers in an NPD to induce the other players to switch to *D* when the game is repeated. There is little or no incentive to cooperate if the goal of collective cooperation cannot be realized: when one's neighbour is seen watering the garden during a drought one is tempted to follow suit. Now if a fixed proportion of the population has a propensity to choose *D*, then the probability of one or more of these "bad apples" turning up in an NPD increases with the size of the group: it is least likely in a small group and most likely in a large group. As group size increases, so does the probability that one or more "bad apples" will turn up and spoil things for everyone by frustrating the goal of collective cooperation. In spite of its superficial plausibility, this theory is severely undermined by the findings of Hamburger (1977) mentioned earlier and by other experiments in which subjects have been pitted against stooge opponents. In these experiments the number of genuine subjects—and therefore the number of hypothetical "bad apples"—is the same in the small and the large groups, yet the larger groups still elicit less cooperation from the genuine subjects. But the bad apple theory is not decisively refuted by these findings. It is possible to argue that subjects act on the *assumption* that there are more likely to be "bad apples" in larger groups, and this assumption might function as a self-fulfilling prophecy.

A second explanation for the decline in cooperation with increasing group size centres on the relatively limited *degree of interpersonal control* that is possible in larger groups (Hamburger, 1979, p. 243). Consider the tit-for-tat strategy, for example: this is a fairly effective method of eliciting cooperation from an opponent in a two-person PDG. But in a large group such a strategy cannot generally be put into effect: it is impossible to choose in such a way as to reciprocate the others' previous choices when some have chosen *C* and others *D*. If we assume that one of the reasons why players choose *C* in repeated two-person PDGs is to exercise interpersonal control, then the relatively smaller proportion of *C* choices in multi-person groups immediately becomes intelligible. This theory accounts satisfactorily for the relatively smaller proportion of *C* choices in multi-person as compared with two-person games. But it does not convincingly explain the differences found in multi-person groups of different sizes: a player has scarcely any more scope for interpersonal control in a three-person than a seven-person NPD, for example.

The third and most satisfactory explanation is the *de-individuation* theory originally proposed by Hamburger, Guyer, and Fox (1975). De-individuation occurs when personal identity and accountability are submerged in a group; in these circumstances individuals become less inhibited about behaving in selfish or antisocial ways. The standard examples of de-individuation in social psychology are taken from instances of mob behaviour. Hamburger, Guyer, and Fox "make the simple assumption that deindividuation in a group increases as the size of the group increases since, everything else being equal, an individual appears more anonymous in a larger group than in a smaller one" (p. 524). Identity and accountability are unavoidable in a two-person game since in that limiting case knowledge that someone has defected necessarily implies knowledge of who it was.

Fox and Guyer (1978) tested the de-individuation theory directly in a four-person NPD experiment in which varying degrees of anonymity were allowed to the players. Some groups of subjects exchanged names and background information while others did not, and the subjects' choices were either public or anonymous. The results revealed significantly lower levels of cooperation in the anonymous conditions as predicted by the de-individuation theory. Similar findings have been reported by Jerdee and Rosen (1974). In an earlier three-person NPD experiment in which the subjects' choices were always anonymous, Kahan (1973) found extremely low levels of cooperation: "The choices made by the individual players were shown to have been made with apparently no regard for the choices of the other two players in the game" (p. 124). All of these findings appear to corroborate the theory of de-individuation.

The effects of opportunities for communication on cooperation in the NPD have been examined by several investigators and, as in the PDG, they have usually been found to promote cooperation to a limited degree. Some commentators have expressed surprise that communication does not generate extremely high levels of cooperation, but since an individual benefits by choosing D in spite of any (non-binding) agreements that might have been reached, the problem is really to explain why communication leads to *any* increase in cooperation. Two possibilities suggest themselves. First, communication may help to foster interpersonal trust and to allay the cooperatively intentioned players' fears that "bad apples" may be present in the group; and secondly, communication is bound to enhance identity and accountability. In either case increased cooperation can be predicted for reasons outlined earlier.

The findings of Mintz (1951) on the effects of communication were outlined above. Two other major investigations in this area are those of Caldwell (1976) and Dawes, McTavish, and Shaklee (1977).

Caldwell's (1976) experiment was based on a five-person NPD in which the subjects were either allowed to communicate with one another and, in some cases, to penalize defectors, or forbidden to communicate. The effect of mere communication was statistically non-significant, but when the subjects were permitted not only to communicate but also to penalize defectors by deducting points from their payoffs, a significant increase in cooperation was observed. Caldwell describes this as a "communication effect" but it amounts, in fact, to an effect of altering the strategic structure of the game by decreasing the payoffs for choosing D. Caldwell admits that "sanctions may have, in effect, changed the nature of the payoff matrix" (p. 278), but he should have said that they *did* change the matrix. No clear conclusion about the effects of communication is therefore possible on the basis of this study.

The carefully controlled study by Dawes, McTavish, and Shaklee (1977) produced unambiguous results concerning communication effects. Subjects were assigned to groups varying in size from five to eight, and the effects of four levels of communication were examined: no communication; communication only about matters irrelevant to the game; relevant communication; and relevant communication plus non-binding public announcements of intended choices before each trial. The first two conditions yielded cooperation rates of 30 per cent and 32 per cent respectively, while the last two yielded rates of 72 per cent and 71 per cent. The effect of relevant communication was highly significant. It is clear from these results that it is discussion of the dilemma rather than merely getting to know the other players through irrelevant chatter that enhances cooperation in the NPD, and also that non-binding announcements of intentions have no noticeable effects. These conclusions harmonize perfectly with the theoretical analysis of the effects of communication in terms of accountability and identification of potential "bad apples" outlined earlier.

Concerning attribution effects, several studies of the influence of the perceived intentions of others on a player's choices are worth discussing briefly. The findings in this area bear on Kelley and Stahelski's (1970c) *triangle hypothesis* discussed at length in Section 7.7. Briefly, there are hypothesized to be two classes of people: those who are habitually cooperative and who learn through experience that others vary in cooperativeness, and those who are habitually competitive and whose erroneous attributions of competitive intentions to others are unlikely to be corrected through experience because their competitive behaviour tends to elicit competitive responses from others.

Tyszka and Grzelak (1976), Alcock and Mansell (1977), Dawes, McTavish, and Shaklee (1977), and Marwell and Ames (1979) have all reported a strong, positive relationship between propensity to cooperate and attribution of cooperative intentions to others. In the second of these studies, for example, defectors predicted 24 per cent cooperation from the other players while cooperators predicted 68 per cent cooperation, and this difference was highly significant. In the Dawes *et al.* experiment defectors predicted four times as much defection from other players as was predicted by cooperators, and the correlation between cooperative choices and attributions of cooperative intentions to others was about 0.60. These findings offer some indirect support for the triangle hypothesis although they can be explained quite easily without it. The triangle hypothesis was conceived in the context of two-person interactions and cannot be generalized without a certain awkwardness to the multi-person context. A direct test in groups varying in size from 10 to 15 (Kelley and Grzelak, 1972) produced ambiguous results, but led to the interesting incidental discovery that cooperators usually have a better understanding of the strategic structure of the game than defectors.

Virtually all published NPD experiments have been based on purely abstract games. Alcock and Mansell (1977), however, reported three experiments in which a slightly more lifelike context was given to the subjects' choices. In all three experiments a conventional numerical payoff matrix was used, but the subjects were told, in addition, that the game was "a simulation of animal population growth under conditions of scarce resources" (p. 447). They were given a verbal account of the "tragedy of the commons" and were assigned the roles of cattle farmers. Their choices were labelled "Add" (an animal to the pasture) and "Not Add". In free-play conditions the level of cooperation was about 30 per cent, which is certainly no higher than might have been expected in a conventional abstract version of the same game. When subjects were provided with false feedback about the choices of the other players on previous trials, whether the others had allegedly chosen highly cooperatively or highly competitively, little effect was observed on the subjects' own choices. No attempt was made in these three experiments, or in any other experiments using the NPD, to compare the behaviour of subjects in abstract and structurally equivalent lifelike decision contexts. A new experiment devoted to this comparison is reported in Section 9.5.

9.5   Experiment IV: Abstract and Lifelike N-Person Prisoner's Dilemmas

According to Davis, Laughlin, and Komorita (1976) "the N-person case (NPD) has greater generality and applicability to real-life situations" than the PDG, and "it is reasonable to assume that empirical studies of the NPD may yield valuable insights into the nature of cooperative behavior in such situations" (p. 520). But Dawes (1980) has expressed the contrary view that NPD experiments "are lousy *simulations* of the social dilemmas with which most of us are concerned. . . . Findings about how small groups of students behave in contrived situations cannot be generalized to statements about how to save the world" (p. 188).

The experiment reported in this section (Colman, 1979a, Experiment IV) was designed to compare behaviour in a conventional abstract NPD with behaviour in a structurally equivalent simulation of a lifelike social dilemma. Its main purpose was to examine the extent to which a lifelike decision context influences the choices of subjects in strategic interactions with the structural properties of the NPD. A secondary aim was to examine the effects of incentives on NPD choice behaviour.

In the matrix (M) condition the subjects played a conventional three-person NPD for points on the basis of a purely numerical table of payoffs. In the positive incentive (PI) condition monetary incentives were introduced, and in the negative incentive (NI) condition the subjects played to conserve as much as possible of a monetary stake given to them at the start of the experiment. These three treatment conditions were entirely abstract, and their payoff structures were equivalent at an interval-scale level of measurement. In the lifelike simulation (S) condition the subjects played the roles of three Ministers of Economics of rival oil-producing nations deciding between restricted production and full production of oil. The payoffs in this condition were imaginary profits, and they bore an interval-scale correspondence to the payoffs in the other three conditions. Any differences between behaviour in the four treatment conditions can be attributed to strategically (and logically) irrelevant psychological differences between the decision contexts.

The major hypothesis was that the S condition would elicit less cooperation from the subjects than the other (abstract) conditions. There are reasons for assuming that a decision context based on business rivalry tends, in our culture, to engage values of competitiveness, and the results of Experiments II and III (see Sections 7.9 and 7.10) tend to confirm this assumption. A second hypothesis was that higher levels of cooperation would emerge from the incentive conditions (PI and NI) than from the non-incentive conditions, since the subjects might feel more strongly motivated to cooperate when something concrete is at stake. Thirdly, it was hypothesized that the psychological differences between playing to maximize gain and playing to minimize loss would be manifested in some difference between behaviour in the PI and NI conditions. A fourth hypothesis, based on theoretical considerations explained in Section 8.7 and findings from two-person PDG research, was that cooperation would decline over trial blocks.

Method

The subjects were 120 undergraduate students (66 males and 54 females) at the University of Leicester, randomly assigned to treatment conditions in groups of three. The procedure was similar to that used in Experiments II and III apart from the modifications necessitated by the different payoff structure of the game and the nature of the lifelike simulation used in the S condition. In every condition the game was repeated 30 times with each triad of players.

Matrix 9.1

Number Choosing L	Number Choosing R	Payoff to Each L-Chooser	Payoff to Each R-Chooser
3	0	3	—
2	1	2	4
1	2	1	3
0	3	—	2

The simplest possible payoff structure was used; it is shown in Matrix 9.1 in the form used in the M condition. In the abstract conditions (PI, NI, and M) detailed typewritten instructions were presented to the subjects explaining the rules of the game and its payoff structure. In the PI condition the payoffs in Matrix 9.1 represented the numbers of half pence that each subject could earn on a single trial. In the NI condition each subject was given 120 half pence at the start of the experiment and lost between zero and four of them on each trial (four units were subtracted from each of the payoffs shown in Matrix 9.1). In the M condition the subjects played simply for points. In the S condition they played for imaginary financial profits according to the following instructions presented to them in typewritten form:

> Your decisions will be based on the following hypothetical situation. You are the Minister of Economics of one of the leading oil exporting countries. The other people in your group are the ministers representing the other leading oil exporters. The decision facing you at the start of each financial year is whether to adopt a policy of restricted oil production or whether to go in for full production, and your *sole objective* is to maximize the revenue your own country will receive from foreign sales of oil; all other considerations are irrelevant to you.
>
> The following facts are known to you and to all the other members of the group. A policy of restricted oil production, provided it is adhered to by all the group members, will result in the price of oil on world markets being kept high for that financial year, and each country will receive £3m in foreign revenue. If, however, one of the ministers opts for full production while the other two restrict production, his country will sell more oil although, because of the effects of supply and demand, the price of oil on world markets will fall somewhat, and this fall will affect all the members of the group. The net effect of this will be that the country opting for full production will earn £4m in foreign revenue, while the other two will be reduced to £2m each. If two of the three countries go in for full production, the corresponding fall in world oil prices will be greater, and they will each earn £3m, while the third country's revenue will be reduced to £1m. If, finally, all three countries go in

for full production, they will each earn only £2m for the financial year. The net effect of these considerations is summarised below.

(a) All three restrict production; each earns £3m.
(b) One country only chooses full production; that country earns £4m, the other two get £2m each.
(c) Two countries choose full production; they each earn £3m, the third earns £1m.
(d) All three choose full production; each earns £2m.

In this treatment condition, RESTRICTED PRODUCTION corresponds to the L option in the abstract conditions, and these are the cooperative (C) option in the underlying NPD. The defecting (D) option is represented by FULL PRODUCTION or R. Payoffs in the PI, M, and S conditions are equivalent on a ratio scale, and payoffs in the NI condition are equivalent to those in the other conditions on an interval scale.

When the experimenter had satisfied himself that all three subjects in a group understood the task fully, 30 trials were played without any communication between players. After each trial the subjects indicated their degrees of satisfaction with the outcome of that particular trial on a five-point rating scale from "Very displeased" to "Very pleased".

Results

The mean numbers of cooperative choices per group in each block of 10 trials are displayed in Table 9.1. The overall grand mean is 8.20, which (bearing in mind that there were three subjects in each group) indicates that slightly more than 27 per cent of the choices were cooperative. Compared with previous NPD experiments this figure is rather low, but the rate of cooperation in the abstract conditions, which were similar to the decision contexts used by previous investigators, is reasonably typical.

Table 9.1   Mean number of C choices per group ($N = 40$)

Treatment condition	Trial block 1	Trial block 2	Trial block 3
PI	10.50	6.00	4.40
NI	12.20	10.80	7.50
M	12.90	9.90	7.40
S	7.90	4.50	4.40

Two significant main effects emerged from an Analysis of Variance, one due to treatment conditions ($F(3, 36) = 5.02, p < 0.01$) and the other due to trial blocks ($F(2, 72) = 53.93, p < 0.001$). *A posteriori* Tukey tests showed that the frequency of C choices in the S condition was significantly ($p < 0.05$) less than the frequencies in the NI and M conditions, and that cooperation declined significantly ($p < 0.05$) from Trial block 1 to Trial block 2 and from Trial block 2 to Trial block 3.

Product-moment correlations between payoffs and satisfaction ratings on corresponding trials were calculated separately for each subject. The means are as follows: PI: $r = 0.78$, NI: $r = 0.73$, M: $r = 0.77$, S: $r = 0.76$. The grand mean is $r = 0.76$, which indicates that approximately 58 per cent of the variance in satisfaction ratings is accounted for by the explicit payoffs. The differences between the correlations in the four treatment conditions are not significant. A subsidiary analysis of the cooperative choices of 20 subjects (five from each treatment condition) who had manifested extremely high payoff–satisfaction correlations, ranging from $r = 0.86$ to $r = 1.00$ with a median of $r = 0.95$, revealed a pattern of results similar to the one found in the main analysis.

Most of the subjects apparently took the task very seriously indeed. In response to post-experimental questioning, one subject in the PI condition described her fellow group members as "a pair of —!", and epithets like "bastards" and "capitalist bastards" were quite common. A subject in the NI condition wrote: "I found the exercise quite exciting, and after a while began to take losses perhaps too seriously". Another subject in this condition wrote: "Green [the label used to identify one of the other group members] shouldn't be in a university. He should be in an E.S.N. school".

Strongly coloured remarks were least common in the M condition as might be expected, but were most common in the S condition rather than in the incentive conditions. Comments about other group members were classified (blind) as positive ("sensible", "reasonable", etc.) or negative ("uncooperative", "silly", "bastards", etc.). The ratio of negative to positive comments was $17:0$ in the S condition and $35:17$ in the abstract (PI, NI, and M) conditions, and this difference was significant $p < 0.007$, two-tailed z test). One subject in the S condition confessed: "I would not make a very good Minister of Economics", and another reflected: "I suppose I'm either an idealist or a sucker". A third subject took a more self-assertive line: "Knowing that some fool would be immovable, perhaps I should just have made sure I'd get the better of them".

Discussion

The results of this experiment corroborate and extend those of Experiments II and III described earlier in this volume. The major hypothesis, that the lifelike decision context would elicit fewer cooperative choices from the subjects than the abstract decision contexts, was strongly confirmed. This was probably due to the fact that the lifelike simulation involved decision making in a hypothetical business context in which prevailing cultural values encourage competitive behaviour. It shows that the reults of NPD experiments based on abstract games cannot be generalized to lifelike situations in which cultural factors play an important part. This is the same conclusion that emerged from the results of experiments on two-person Prisoner's Dilemma and Chicken games described in Sections 7.9 and 7.10.

The hypotheses about the effects of monetary incentives were not confirmed. It is apparent that, at least when relatively small amounts of money are used, they do not have any clear-cut effects on behaviour in the NPD.

The hypothesis that cooperation would decline over trial blocks was strongly confirmed: the frequency of cooperative choices declined significantly from Trial block 1 to Trial block 2, and from Trial block 2 to Trial block 3. The decline was found in all treatment conditions, and probably reflects the inexorable movement towards the equilibrium point of the game (see Section 8.7). An initial reservoir of goodwill, a naïve lack of understanding of human nature, or perhaps a failure to grasp the strategic structure of the game was gradually superseded by an awareness that individual interests are best served by defecting choices.

The payoff–satisfaction correlations revealed that many subjects departed from the explicit payoff structure by introducing extraneous utilities. The mean correlation of 0.76 indicated that slightly less than 60 per cent of the variance in subjective preferences was attributable to the explicit payoffs, and the correlations were not significantly different from one treatment condition to another. The use of abstract decision contexts is based largely on the assumption that this allows maximum control over extraneous utilities, but the data appear to refute this assumption: subjects do not adhere any more closely to the payoff structure when the game is presented abstractly than they do in the lifelike simulations used in this experiment or in Experiments II and III. The exceptionally high level of ego-involvement in the lifelike simulation provides one strong argument in favour of lifelike games in future research. Ego-involvement was noticeably less in the abstract games, even in conditions involving financial incentives (which were absent from the lifelike simulation).

The results of this experiment cast doubt on the ecological validity of conventional NPD experiments, but two important caveats need to be mentioned. In the first place, it should be borne in mind that no direct comparison was made between behaviour in an abstract experimental game and a strategically equivalent naturally occurring interaction. It is probably impossible to make such a comparison with sufficient control to enable interesting conclusions to be drawn. The lifelike simulation used in this experiment was—deliberately and unavoidably—artificial and contrived in several respects. But it seems reasonable to suppose that, compared to the conventional abstract versions of the game, the lifelike simulation provided a more reliable indication of how the subjects would behave in certain NPD-type strategic interactions that might crop up in their everyday lives. The lifelike simulation serves as an intermediate case between a purely abstract game and a "real world" strategic interaction. It seems reasonable to assume that behaviour in an ego-involving lifelike simulation is more true to life than behaviour in an abstract game.

Secondly, it is not known to what extent and in what ways the findings reported here are contingent upon the particular lifelike simulation chosen for investigation. It was constructed in a manner explicitly calculated to encourage competitiveness, and the subjects responded to it by exhibiting highly competitive behaviour as predicted. It is likely that a context in which the prevailing cultural value favours cooperativeness would have elicited highly cooperative behaviour, but this hypothesis needs to be tested empirically. It is simply not known what cultural values are engaged by abstract games or what interpretations they suggest to experimental subjects.

9.6 Summary

The chapter opened with a comment on the decline of the Prisoner's Dilemma game and the rise of the N-Person Prisoner's Dilemma and other multi-person games as objects of empirical research during the 1970s. Research on coalition formation in cooperative multi-person games, which has a longer history, was reviewed in Section 9.2. Orthodox game theory, based on the characteristic function, does not make specific predictions about which coalitions will form, but it shows which are potentially winning combinations. Minimal winning coalition theory, also known as the size principle, predicts that one of the smallest winning combinations will form. Experimental evidence and historical records of political coalitions tend to corroborate the theory. Minimum resource theory is generally more specific in predicting that a winning coalition whose combined weight is least will form. This theory has fared remarkably well under experimental testing in spite of its often counterintuitive predictions, such as the power inversion effect in which the player who controls the most resources turns out to be disadvantaged relative to the apparently weaker players. Utter confusion theory correctly predicts that in certain difficult circumstances coalitions will form more or less at random. None of the theories is successful in predicting how the members of a coalition will agree to distribute the payoff among themselves.

Section 9.3 was devoted to a comprehensive review of published experiments on auction games and related psychological traps. In auction games with many players the bidding invariably exceeds the value of the prize, and in dyads it frequently does so. The subjects often become highly emotional as the process of escalation gets out of control. At a certain "point of no return" economic motives give way to competitive face-saving motives and the bidders usually "go for broke", apparently unable or unwilling to extricate themselves from the spiral of escalation. It was suggested that this phenomenon be called the Macbeth effect. Closely related to auction games are psychological traps of the type that occur, for example, when a person has invested valuable time waiting for a bus which fails to arrive and feels increasingly reluctant to leave the bus stop. Experiments have focused on factors affecting entry into and inability to escape from entrapment situations.

In Section 9.4 experiments on the N-Person Prisoner's Dilemma were critically reviewed. Cooperation has been found to decline as the number of players increases, and several theories (the "bad apple" theory, decreased opportunities for interpersonal control, and de-individuation) have been proposed to explain it. Opportunities for communication tend to promote cooperation. A strong relationship has been found between cooperation and attributions of cooperative intentions to others. Section 9.5 was devoted to a new experiment on behaviour in abstract and lifelike N-Person Prisoner's Dilemmas. The results showed that a lifelike simulation of business rivalry between three oil-producing nations elicited much less cooperation than structurally equivalent abstract games with or without monetary incentives. Some subjects departed from the explicit payoff structure by introducing extraneous utilities, but no more so in the lifelike than the abstract games. The major conclusion, as in the new experiments described in earlier chapters, is that experiments based on abstract games have limited ecological validity.

Applications

10

Sincere Voting and Collective Choice Theory

10.1 Background

IN MODERN societies, multi-person games are often governed by formal rules designed to ensure that the outcomes are "fair" and "democratic". Formal rules of this kind typically apply to games whose outcomes are collective choices based upon the individual preferences of the players. Committees, for example, often have to choose a single course of action from among a diverse set of proposals favoured by different members. Any method of resolving differences of opinion into a single choice from a set of alternatives is called a *collective choice rule*, and those that have a claim to being "fair" and "democratic" are usually voting procedures of one kind or another. Voting procedures are used by electorates for choosing political representatives, by legislatures for choosing laws, by colleges of cardinals for choosing popes, and by juries, boards of directors in industry, groups of shareholders, trade unions, and many other kinds of decision making bodies.

A number of paradoxes associated with apparently reasonable voting procedures were discovered by mathematicians and physicists of the French Enlightenment. The first major contribution was made by Borda (1781), who was followed by Condorcet (1785). During the nineteenth century, the Reverend C. L. Dodgson (1876), better remembered as Lewis Carroll, independently rediscovered some of the paradoxes. The subsequent historical development of collective choice theory has been traced by Black (1958) and Riker (1961). More recent literature in this area has been surveyed by Sen (1970), Fishburn (1973a), and Kelly (1978).

A tacit assumption underlying the bulk of collective choice theory is that voting is *sincere*. This is an assumption about the strategies that the players—in this case the voters—will adopt in every voting game. In any genuine ballot, a voter has at least two pure strategies, otherwise there would be no real element of choice. Voting is said to be sincere if the voters always vote in favour of the alternatives they consider best. A voter who prefers an alternative x to another y, and prefers y to a third z, votes sincerely if he or she votes for x whenever it is available in a ballot, and if the choice is between y and z, votes for y. In the terminology of game theory, sincere voting corresponds to choosing the maximax strategy (see Chapter 2). A sincere voter in effect selects the strategy that might produce the best of the best possible outcomes—from this voter's point of view—given some conceivable patterns of strategy choices by the other voters. Many authors define sincere voting loosely as "voting strictly according to one's preferences"; but this is misleading because insincere voting sometimes accords better with one's preferences than sincere voting. A maximax strategy is not necessarily the most sensible choice in a multi-person game whether it is a voting game or not. In many circumstances voters evidently do nevertheless vote sincerely without even considering the consequences of other strategy choices. In this chapter sincerity will be assumed; the implications of insincere or "strategic" voting are spelled out in Chapter 11.

In the following two sections, the formal structure of collective choice theory is outlined. Three common voting procedures are examined in Section 10.4. Sections 10.5 and 10.6 are devoted to the most famous paradox of voting, namely Condorcet's paradox, and in Section 10.7 Arrow's impossibility theorem, which is based on the paradox, is discussed. A less well known paradox of voting, the Borda effect, is introduced in Section 10.8, and the chapter is summarized in Section 10.9.

10.2   Alternatives, Voters, Preferences

The following hypothetical example will help to sharpen the primitive terms of the theory. A committee is faced with the problem of choosing one of three candidates to fill a vacant political office. One of the candidates, Leftwich, is known to hold left-wing views; another, Middleton, is politically moderate; the third, Rightsman, is extremely right-wing. There are three opinion blocs among the committee members. The "leftist" committee members (obviously) prefer Leftwich to Middleton and Middleton to Rightsman: their preference scale may be written compactly as LMR. The preference scale of the moderate committee members is MLR, and that of the "rightists" is obviously RML. These three opinion blocs are the only ones represented on the committee, and they are of approximately equal voting strength; none commands more than half the total number of votes in the committee, so a combination of any two can outvote the third.

In this idealized example, there are three *alternatives* from among which a collective choice has to be made: L (Leftwich), M (Middleton), and R (Rightsman). The set of *voters* can be defined in two different ways: either as the members of the committee, or as the distinct opinion blocs. In the latter case we need consider only three effectively distinguishable voters: leftist, moderate, and rightist; this would be possible even if the choice were being made by an entire electorate with just three opinion blocs, none of which commands more than half of the votes. Finally, each voter has *preferences* among the alternatives which we can represent by means of an ordered triple of initials. The assumption here is that each voter is capable, if requested to do so, of ranking the alternatives in order of preference from "best" to "worst".

10.3   Axioms Concerning Individual Preferences

The primitive terms outlined in the previous section can be formalized by means of a mathematically defined family of *strong ordering relations* between pairs of alternatives. For technical reasons, and also to take account of certain basic assumptions about human rationality, two important restrictions are placed on these relations. The first is *completeness* or *connectedness*: it is assumed that for every pair of alternatives x, y, each voter either prefers x to y or y to x but not both. This assumption is often relaxed in order to allow a voter to express indifference between a pair of alternatives, but this leads to complications which are not needed here. The second assumption is *transitivity*: if x, y, and z are three alternatives, and if a voter prefers x to y and y to z, then he or she must also prefer x to z. The strong ordering relation P ("is preferred to") is defined for each voter and for each pair of alternatives by incorporating the assumptions into the model as axioms:

(i) *Completeness:* for every voter and for every pair of alternatives x, y, either xPy or yPx but not both.

(ii) *Transitivity:* for every voter and for every triple of alternatives x, y, z, if xPy and yPz, then xPz.

The first axiom is relatively uncontroversial; it simply ensures that the preference relation is defined for all voters over all pairs of alternatives. The second, however, implies an important assumption about human rationality which is not as innocent as collective choice theorists usually seem to think. Do rational human beings ever prefer x to y, y to z, and z to x? Not if we define rationality according to the axiom of transitivity! But an abstract model, if it is to be useful, must correspond to everyday experience, so the relevant question is this: Can plausible counterexamples to the axiom of transitivity be found? It turns out to be extremely difficult to imagine them. Consider the following, however: Someone prefers an office on the fifth floor to one on the ground floor because of the better view, and, for the same reason, prefers the tenth floor to the fifth. But this same person prefers the ground floor to the tenth floor because of the inconvenience of climbing ten flights of stairs when the lifts are out of order! (This is a sharpened version of a counterexample originally produced by Hamburger, 1979, p. 37.)

Empirical evidence (see Edwards & Tversky, 1967, pp. 44–47, 77–78) is consistent with the view that *in most circumstances* intransitivities in individual preferences arise only from carelessness when people are required to make numerous unimportant comparisons. When the intransitivities are pointed out, the decision makers usually proceed to eliminate them (Niemi & Riker, 1976). They can, of course, be avoided altogether by requiring each decision maker to arrange all of the alternatives in a single rank order. Virtually the entire edifice of collective choice theory rests upon the axiom of transitivity. I have the feeling that more attention deserves to be devoted to examining it empirically and, more important still, to working out the theoretical consequences of abandoning it.

The ingredients—the primitive terms and the axioms—of the simple model of collective choice involving Leftwich, Middleton, and Rightsman are now assembled. The voters' preferences, given Axioms (i) and (ii), can be summarized as shown in Table 10.1.

Table 10.1

	Voters (or voting blocs)		
	Leftist	*Moderate*	*Rightist*
Most preferred alternative	L	M	R
Second ranked alternative	M	L	M
Least preferred alternative	R	R	L

10.4   Voting Procedures

A very common voting procedure used, for example, in the legislatures of West Germany, Denmark, and Norway, and in the Council of Europe (Bjurulf and Niemi, 1981) is the *successive* procedure. A ballot is held on each alternative in turn; the voters vote in favour or against in each case, until a majority votes in favour of one of the alternatives. If none attracts a majority of votes before just two alternatives remain, then a final ballot between these two, with the voters voting in favour of one or the other, decides the issue. In the example of the committee, the voters might first be required to vote in favour or against L (Leftwich); if a majority votes for L then a collective decision has been reached and further balloting is unnecessary, but if a majority votes against L (i.e. in favour of "M or R", which means the same as "not L") then a second and final ballot is held between M and R. The successive voting procedure can begin with a ballot on any one of the alternatives and, as we shall see, different orders of voting do not necessarily generate the same collective choice from a fixed patterns of individual preferences.

If L is presented first, then the sincere outcome is as follows:

First ballot: In favour of L (Leftwich), leftists;
                Against L, moderates and rightists.
                Result: M or R. Therefore
Second ballot: In favour of M (Middleton), leftists and moderates;
                In favour of R (Rightsman), rightists.
                Final Result: M.

The collective choice is Middleton, the alternative which the moderate voters consider best and the leftists and rightists consider second best. If, on the other hand, M is presented first, the outcome is:

First ballot: In favour of M, moderates;
                Against M, leftists and rightists.
                Result: L or R. Therefore
Second ballot: In favour of L, leftists and moderates;
                In favour of R, rightists.
                Final Result: L

With this order of voting, Leftwich wins! It is easily verified that if Rightsman had been presented first, Middleton would have won. An alternative presented in an early ballot under the successive procedure is at a relative disadvantage if voting is sincere. An interesting consequence of the above analysis is this: if the voters knew the preferences of the others, that is to say, if the voting game were one of complete information (see Chapter 1), and if they were permitted to choose the order in which the votes should be taken, the leftists would favour presenting Middleton first, thus ensuring that Leftwich would be the final choice as shown in the second example. The moderates and rightists, on the other hand, would favour presenting Leftwich or Rightsman first, because they would prefer the final choice to be Middleton rather than Leftwich as in the first example. There are only two effectively distinct voting orders: "M first" and "L or R first". A single ballot would decide this procedural issue in favour of "L or R first" because only the leftists would favour the other order. The final result of balloting in the order favoured by the majority would be the choice of Middleton.

Procedural votes of the kind described above are permitted in some decision making bodies but not in others. They are permitted in the Swedish parliament or riksdag (Rustow, 1955) and in British Labour Party Conferences, for example. *The Times* (24 January 1981) reported that

> a procedural strategy adopted by the moderate leaders of the Amalgamated Union of Engineering Workers (AUEW) could threaten the precariously balanced centre–left coalition supporting the compromise formula [for electing the party leader] favoured by Mr Michael Foot . . . by getting delegates to reject the standing orders committee recommendation for debating procedure (p. 2).

In the event, the order of voting led to the "compromise formula" being eliminated before the final ballot as expected. But the AUEW was hoist with its own petard, because a left-wing motion, granting trade unions the largest say in the election of the party leader, defeated a right-wing motion on the final ballot (*The Times*, 26 January 1981).

A second voting procedure which is often used in legislative bodies is the *amendment* procedure. It is common in Great Britain and all of her former colonies, including the United States, and also in Sweden and Finland (Bjurulf and Niemi, 1981). A substantive motion is tabled and an amendment (or a number of amendments) is proposed. The first ballot decides in favour or against amending the motion, and the final ballot determines whether the (possibly amended) motion is passed or rejected. Rejecting the motion on the final ballot is equivalent to choosing a third (default) alternative. Different orders of voting are possible with this procedure as well, since any of the alternatives might be treated as the substantive motion and any other(s) as the amendment(s). In the committee example, the substantive motion might be "That Rightsman be appointed to the vacant post" and the amendment might be "That the word 'Rightsman' be replaced by 'Middleton'". Leftwich is then the default alternative, assuming that one of the three has to be appointed. The first ballot is then on whether to amend the motion:

First ballot: In favour of M (the Middleton amendment), leftists and moderates;
In favour of R (Rightsman, the substantive motion), rightists.
Result: M.

Second ballot: In favour of M, moderates and rightists;
Against M (i.e. in favour of L), leftists.
Final Result: M.

It can easily be checked that with the postulated preferences of the voters the same final result emerges from any order of voting under the amendment procedure. But this is not always the case. Had the preference scales of the voters been LMR, MRL, and RLM, then any of the alternatives could emerge as the final choice depending on voting order. This is an example of Condorcet's paradox, to which we shall return later in this chapter. Riker and Ordeshook (1973), Miller (1977) and Bjurulf and Niemi (1981) have discussed the conditions under which voting orders influence final choices in various sequential procedures including successive and amendment voting. They have also outlined the strategies that voters might be expected to adopt in voting on procedural issues in these circumstances.

The most common voting procedure in committees and elections is undoubtedly *plurality* voting: all of the alternatives are pitted against one another in a single ballot, and the one that receives the largest number of votes is declared the winner. In order to determine which alternative would be chosen by plurality voting in the committee example, it is necessary to make additional assumptions about the voting strengths of the opinion blocs or voters.

Most voting procedures include special rules to resolve ties and to ensure that a decisive result is always reached. In committees, for example, the chairman may have a casting vote in addition to his ordinary deliberative vote, to be used in the event of a tie on any ballot. (This is common in most English-speaking countries, though not in the United States.) In the committee example, we may arbitrarily assume that either the leftists are the largest of the three voting blocs, or the largest voting blocs are of exactly equal size but the additional casting vote belongs to one of the leftists. With these assumptions it is obvious that the result of sincere voting is the choice of Leftwich.

Three common voting procedures have been examined in this section. With the postulated preferences and voting strengths, and assuming that all voters are sincere, the successive procedure results in the choice of Leftwich or Middleton depending on the order of voting. The amendment procedure always leads to the choice of Middleton. The plurality procedure ensures that Leftwich is chosen. Given a choice of voting procedures, the leftists would favour the successive procedure with Middleton presented first, or the plurality procedure, since these procedures ensure the choice of their favourite candidate. Both the moderates and the rightists, on the other hand, would favour the successive procedure with Leftwich or Rightsman presented first, or the amendment procedure, because Middleton is then bound to be the final choice, and they prefer Middleton to Leftwich. Choosing one of the voting procedures "democratically" presents problems, however, because a voting procedure of some kind would have to be used in order to make this choice, and opinions might differ on the procedure to be used for choosing a procedure. There is an obvious infinite regress here: one is confronted with the prospect of a committee bogged down by an unending series of points of order, unable to take any votes at all. Such is the price of sincerity, for it will be shown in Section 11.8 that this problem is partly solved (in a manner consistent with the will of the majority) by insincere voting.

10.5   Condorcet's Paradox

The Marquis de Condorcet (1785) seems to have been the first to discover a voting paradox that lies at the centre of modern collective choice theory. Its simplest manifestation is in a group of three voters choosing among three alternatives A, B, and C. Assuming as before that the voters' preference scales are complete and transitive, consider the possible pattern shown in Table 10.2. This pattern is known as a *Latin square*; it has the property that each letter appears exactly once in each row and in each column. In other words, each alternative appears exactly once in each position—first, second, third—in the three preference scales. This means that no two voters agree about which alternative is best, which is second best, or which is worst. The only other Latin square which can be formed from three alternatives A, B, and C is based on the preference scales ACB, BAC, CBA, and this pattern also generates Condorcet's paradox.

Table 10.2

	Voters		
	1	*2*	*3*
Most preferred alternative	A	B	C
Second ranked alternative	B	C	A
Least preferred alternative	C	A	B

The paradox emerges if we assume that the voters are sincere, and then try to determine the collective choice according to majority rule. Two of the three voters (1 and 3) prefer A to B, two (1 and 2) prefer B to C, but two (2 and 3) prefer C to A. Thus a majority of the voters prefer A to B, B to C, and C to A. The collective preference scale of the group cannot be written down in the normal way: a linear ordering of alternatives takes the form $ABCABCA$ . . . which contradicts itself. A set of transitive individual preferences has generated an intransitive collective preference called (for obvious reasons) a *cyclic majority*. The "group mind" can therefore violate a basic axiom of rationality (Axiom (ii) in Section 10.3 above) even if all of its members are perfectly rational. Condorcet's paradox looks harmless enough on the surface, but it has disturbing implications for democratic theory and practice, as we shall see.

Some kind of collective choice rule is needed in order to make a collective decision on the basis of a set of individual preferences. Some collective choice rules are obviously undemocratic. For example "Write each alternative on a separate slip of paper and draw one of them out of a hat at random" is clearly a workable collective choice rule, but it is undemocratic because, in the terminology of collective choice theory, it is *imposed*: the choice does not depend in any way on the preferences of the individuals. An imposed rule may be considered acceptable for lotteries, raffles, pools, and other games of chance, but few would recommend transforming the decisions of legislatures, juries, and other decision making bodies into games of chance. Another workable collective choice rule might be: "Choose the alternative that the President considers best". But this rule is *dictatorial*: the choice depends entirely on the preference of one individual. It is blatantly undemocratic. A rule that seems on the face of it entirely unexceptional is the *method of majority decision*: "Choose an alternative that is preferred to each of the others by a majority of the individuals". This rule has won a special place in the hearts of collective choice theorists because it has the makings of a democratic procedure but also has the interesting property of being unworkable in some circumstances. To be specific, it is impossible to choose an alternative from the cyclic profile of preferences shown above by the method of majority decision because there is no alternative that is preferred to each of the others by a majority.

Condorcet's paradox leads immediately to an elementary impossibility theorem (Colman and Pountney, 1975b). It is this: no collective choice rule ever devised, nor any other rule that could ever be devised in the future, could satisfy the following two conditions of workableness and fairness: (i) an alternative must always be chosen, no matter what the profile of individual preferences; (ii) none of the rejected alternatives must ever be preferred to the chosen alternative by a majority of the voters. The proof is transparently simple: if a collective choice rule satisfies (i) it must make a choice from a pattern of preferences that generates a cyclic majority such as the one shown above. In that case A or B or C would therefore have to be chosen. But A cannot be chosen without violating (ii) because a majority of the voters prefers C to A; similarly B is an unacceptable choice because a majority prefers A to B; and C is ruled out because a majority prefers B to C. No collective choice rule could satisfy both (i) and (ii) with the postulated profile of preferences, therefore none could do so in all cases. This completes the proof.

Some collective choice theorists might consider the second condition less intuitively necessary (or less ethically compelling) for a fair collective choice rule than the conditions used in other impossibility theorems, including Arrow's (1951, 1963) famous theorem discussed below in Section 10.7. But the theorem outlined in the previous paragraph has the virtue of simplicity, and its proof makes no assumption about the sincerity of the voters.

Condorcet's paradox has often been called "the paradox of voting" since Nanson (1882) so named it. This phrase is misleading, however, because a large number of paradoxes of voting have now been discovered. Some others are described in this chapter and in Chapter 11, and further examples are discussed by Fishburn (1974a) and Brams (1976). Condorcet's paradox is, nonetheless, undoubtedly the most important. It is therefore worth demonstrating how it might plausibly occur in an election. The following example will also show that the paradox is not restricted to the simple case of three alternatives and three voters.

Imagine a large group of voters—a constituency in an election, say—containing 30,000 members. Roughly one-third are poor, one-third are in the middle-income bracket, and one-third are rich. The main election issue is taxation: some candidates have promised, if they are elected, to impose new taxes on the poor or the rich or both. All voters are mildly in favour of any new tax that does not apply to their own group, because they believe that they might benefit indirectly from the increased government revenue. But all voters strongly object to a new tax on their own group. Four candidates, *A, B, C,* and *D* contest the election on the following platforms:

A: New taxes on rich and poor.
B: New taxes on poor only.
C: New taxes on rich only.
D: No new taxes.

The preference scale of the 10,000 poor voters is (obviously) *CDAB*. The preferences of the 10,000 middle-income earners between *B* and *C* depend on whether they prefer the rich or the poor to suffer new taxes. Assuming the latter, their preference scale is *ABCD*. The preference scale of the 10,000 rich voters is clearly *BDAC*.

There is something distinctly peculiar about this pattern of preferences. Two out of three voters (the middle-income and rich voters) favour new taxes on the poor, and a similar majority (the middle-income and poor voters) favour new taxes on the rich. Nonetheless, if voting is sincere, the candidate who includes both of these popular proposals in this platform, namely *A*, could be defeated by the one who includes neither, namely *D*, if the other two candidates withdrew before the election took place! This is so because two out of three voters (the poor and the rich) prefer *D* to *A*. If *B* and *C* did not withdraw, *D* would not receive a single vote, however, because *D* is not the favourite candidate of any voter! The trouble arises from the existence of a cyclic majority: a majority prefers *A* to *B*, *B* to *C*, *C* to *D*, and *D* to *A*. Although this example is slightly artificial, it shows how a cyclic majority might plausibly arise in a complex situation with rational individual preferences. For detailed discussions of Condorcet's paradox arising from the combination of issues and platforms, see Downs (1957) and Hillinger (1971).

10.6 Probabilities of Cyclic Majorities

When there are at least three alternatives and at least three voters, cyclic majorities are always possible. It seems reasonable to ask how likely they are, but this question is in fact a very difficult one to answer. It is ultimately an empirical question which needs to be tackled through systematic observations under controlled conditions. But social psychologists, who are perhaps best equipped to carry out such investigations, have not shown any great interest in this problem, in spite of its important political, moral, and practical implications. Some light has, however, been shed on the problem by purely theoretical arguments.

It is possible to calculate theoretical probabilities of cyclic majorities by making certain assumptions. The earliest calculations along these lines were devoted to the simplest case of three alternatives and three voters. The most elementary assumption about the preferences of the voters, and the one which was used in these early calculations, is that of an *equiprobable culture*: this is a profile of preferences that arises if each voter adopts each of the logically possible preference rankings with equal probability. In the three-alternative case, for example, an equiprobable culture is one in which each voter adopts one of the six possible preference rankings *ABC, ACB, BCA, BAC, CAB*, or *CBA* strictly at random; the probability of a particular ranking being chosen by a particular voter is 1/6.

With three voters, the probability of a cyclic majority can be worked out quite simply by combinatorial analysis. We begin by calculating the number of possible profiles of preferences among the voters. The first voter can choose any one of six rankings, and for each choice there remain six for the second voter and six for the third. There are thus $6 \times 6 \times 6 = 216$ possible profiles of preferences. For a cyclic majority to exist, the rankings chosen by the three voters must either be the first, third, and fifth in the set listed in the previous paragraph, or the second, fourth, and sixth. The first voter can choose any one of the six rankings, but for each choice there remain only two for the second voter and one for the third. There are thus $6 \times 2 \times 1 = 12$ ways in which three voters can choose a set of preferences generating a cyclic majority. Of the 216 possible profiles, 12 are cyclic; therefore if all preference rankings are equally likely to be chosen, the probability of a cyclic majority is $12/216 = 0.056$. In other words it may be expected to occur between five and six times in a hundred.

The first to report the above result were Guilbaud (1952) and Black (1958). Similar methods were later used by Garman and Kamien (1968), Niemi and Weisberg (1968), Gleser (1969), DeMeyer and Plott (1970), May (1971), Blin (1973), Kelly (1974) and others (see also Niemi and Weisberg, 1972, Part 3) to calculate the probabilities in equiprobable cultures with more than three alternatives and/or voters. Some of the results are summarized in Table 10.3.

Table 10.3   Probabilities of cyclic majorities in equiprobable cultures

Number of alternatives	Number of voters				
	3	*5*	*7*	...	∞
3	0.056	0.069	0.075	...	0.088
4	0.111	0.139	0.150	...	0.176
5	0.160	0.200	0.215	...	0.251
6	0.202	0.255	0.258	...	0.315
7	0.239	0.299	0.305	...	0.369
⋮	⋮	⋮	⋮	⋮	⋮
∞	1.000	1.000	1.000	1.000	1.000

Two things are particularly noteworthy about these results. First, the probabilities are relatively insensitive to increases in the number of voters. With three alternatives, for example, the probability is never less than 0.056, or about 1 in 18, and it is never more than 0.088, which is still less than 1 in 11. The probabilities are, on the other hand, extremely sensitive to increases in the number of alternatives: as the number of alternatives increases in a group of any size (larger than two), the probability of a cyclic majority rises rapidly towards certainty.

It must be borne in mind when examining the probabilities displayed in Table 10.3 that they are applicable only to equiprobable cultures: every voter is assumed to be indifferent to every alternative to the point of choosing a preference ranking completely at random. In reality this assumption is seldom justified; objective interests and social pressures of various kinds almost always make some alternatives appear preferable to others, at least in the opinions of some voters. One is tempted to say that an equiprobable culture is *prima facie* highly improbable.

A more promising theoretical approach is the attempt to discover the conditions that a profile of preferences must satisfy in order to guarantee either the existence or the non-existence of a cyclic majority. There is one well-known condition which rules out the possibility of a cyclic majority. This condition, known as *single-peakedness* was originally discovered by Black (1948a, 1948b). It is satisfied if all of the voters evaluate the alternatives according to the same single criterion. In a political context, for example, a profile of preferences will be single-peaked if the voters all evaluate the candidates solely according to how left-wing they are, as in the example of Leftwich, Middleton, and Rightsman discussed earlier in this chapter.

A profile of preferences is single-peaked if it is possible to arrange the alternatives in some order along a horizontal axis so that the graph of each voter's preference ranking, with ordinal degrees of preference represented on the vertical axis, has just one peak corresponding to his or her most preferred alternative. If Leftwich, Middleton, and Rightsman are arranged in that order along the horizontal axis, for example, then any voter who evaluates the candidates solely according to the left–right dimension of political views will necessarily have a single-peaked preference curve. A leftist voter's peak will be at the left-hand side of the graph, a moderate's peak will be in the middle, and a rightist's peak will be on the right. If the voters evaluate the candidates according to more than one criterion, on the other hand, the profile of preferences might not be single-peaked. In that case the voters' preference scales might be *LMR*, *MRL*, and *RLM*, for example; there is then no arrangement of alternatives that results in all three curves being single-peaked. The example in Section 10.5 above, in which the voters evaluated the candidates *A*, *B*, *C*, and *D* according to two independent criteria (their policies on taxing the poor and the rich) was specifically designed to violate the single-peakedness condition. In that case, it will be recalled, a cyclic majority emerged. What Black (1948a, 1948b) managed to prove was that, provided the number of voters is odd, the condition of single-peakedness is sufficient to rule out the possibility of a cyclic majority. A more elegant mathematical interpretation of single-peakedness was later given by Coombs (1964, chaps 9 and 19) in terms of his unfolding theory.

Cyclic majorities have not been investigated under laboratory conditions, but a number of plausible empirical examples from real life have been reported. In most cases, of course, the complete preference rankings of the voters are not known; in elections, for example, voters are normally required to indicate their top-ranked alternative only, and even this can be misleading if voting is not sincere. The examples which have been reported in the literature are therefore based upon more or less reasonable conjectures about the preference rankings of the voters. In the absence of more solid empirical evidence they are not, however, without interest. Riker (1965) provided detailed evidence suggesting that Condorcet's paradox has occurred on more than one occasion in the United States Congress, and Bowen (1972) has shown that it has probably occurred quite frequently in Senate roll call votes. Niemi (1970) has given plausible examples from faculty elections in universities. Brams (1976, chap. 2) has summarized a number of empirical examples from the field of politics. Several amusing examples of intransitive relations in sports and games have been outlined by Gardner (1974).

10.7   Arrow's Impossibility Theorem

Arrow (1951) was the first to use Condorcet's paradox to prove that no conceivable collective choice rule can satisfy certain minimal conditions of ethical acceptability or "fairness" and workability. It probably came as a relief to many people when Blau (1957) found an error in Arrow's proof and even produced a counterexample, in the form of a collective choice rule that satisfied all of Arrow's conditions. The relief was short-lived, however, because Arrow (1963) managed to modify his theorem in a way which not only removed the error but also strengthened the conclusion.

Arrow's theorem requires a collective choice rule to be a *social welfare function*. This is a method of choosing, on the basis of a set of individual preference rankings, not merely a single "best" alternative, but a complete and transitive collective ranking of all alternatives from "best" to "worst". The simple plurality voting procedure, for example, is a social welfare function because it allows the alternatives to be ranked in order of collective preference. Arrowian social welfare functions are usually based on weak ordering relations; that is to say, individual and collective preferences are permitted to include ties between pairs of alternatives in addition to strong preferences. Social welfare functions based on strong preferences only can, however, be more simply illustrated. For the case of two alternatives and two individuals, some typical social welfare functions are shown in Table 10.4. The table shows three possible social welfare functions from a much larger set ($2 \times 2 \times 2 \times 2 = 16$) which can be constructed in this simple two-alternative two-person situation. For each possible profile of individual preferences, listed on the left, Functions I, II, and III generate different collective preference rankings. For example if Individual 1's preference ranking is AB, and that of 2 is also AB, Functions I, II, and III generate the collective preference rankings AB, BA, and AB respectively, as shown in the first row of the table. Function I is evidently dictatorial because the collective ranking always reflects the preferences of Individual 1. Function II is imposed: it is completely unresponsive to the preferences of the individuals. Function III is not obviously unfair on the face of it; at least the collective ranking corresponds to the individual rankings when the latter are unanimous. In other words, Function III, unlike Function II, satisfies the weak Pareto condition (see below).

Table 10.4   Individual preference rankings and social welfare functions; two individuals and two alternatives

Individuals		Social welfare functions		
1	*2*	*I*	*II*	*III*
AB	AB	AB	BA	AB
AB	BA	AB	BA	AB
BA	AB	BA	BA	AB
BA	BA	BA	BA	BA

Arrow (1963) assumed that any acceptable social welfare function would have to satisfy the following four apparently mild and uncontroversial conditions:

U: *Unrestricted domain.* The social welfare function must be wide enough in scope to generate a collective preference ranking from any logically possible set of individual preference rankings.

P: *Pareto condition* (weak form). Whenever all of the individuals prefer an alternative x to another y, x must be preferred to y in the collective preference ranking.

I: *Independence of irrelevant alternatives.* The collective ranking of any pair of alternatives x and y must depend solely on the individuals' rankings of these alternatives and not on their rankings of "irrelevant" alternatives.

D: Non-dictatorship. The collective ranking must not be dictated by the preferences of one individual; it must not be the case that whenever a specified individual prefers any alternative x to any other y, x is necessarily preferred to y in the collective ranking regardless of the preferences of the other individuals.

The only condition that requires any elaboration is I. What is intended is this: If the social welfare function determines that Leftwich is collectively preferred to Middleton, this must be based solely on the voters' rankings of Leftwich *vis-à-vis* Middleton. The collective rankings of Leftwich and Middleton must, in other words, be independent of the voters' rankings of irrelevant pairs, such as Leftwich and Rightsman, Middleton and Rightsman, or Rightsman and Hitler. In other words, the collective preference ranking of Leftwich *vis-à-vis* Middleton must remain the same if one or more of the voters change their minds about the relative merits of pairs of irrelevant alternatives, provided that their rankings of Leftwich *vis-à-vis* Middleton remain fixed.

Arrow's (1963) astonishing theorem asserts that no social welfare function can possibly satisfy all four of the conditions U, P, I, and D. The fundamental ideas of the proof can be outlined quite simply. Condition U allows any profile of individual preferences to be postulated; Arrow therefore postulated a profile that generates a cyclic majority *à la* Condorcet's paradox. He then proved that any social welfare function that satisfies P and I necessarily violates D in this case, that is to say it is dictatorial.

A variety of alternative proofs of Arrow's theorem have been devised, and literally scores of related impossibility theorems have been proved. Kelly (1978) has provided a useful "field guide" to this difficult body of literature. People who are unfamiliar with recent work on collective choice theory often believe that the importance of Arrow's theorem can be denied by criticizing the intuitive plausibility of one of his conditions. It has been argued many times, for example, that Condition I is not really absolutely necessary for a fair and democratic collective choice rule. Kelly has pointed out, however, that *"for each of Arrow's conditions, there is now an impossibility theorem not employing that condition"* (p. 3, italics in original). The general conclusion is no longer even restricted to social welfare functions; impossibility theorems have been proved for all types of collective choice rules including all voting procedures which are designed merely to choose a single "best" alternative (Blair *et al.*, 1976). The conclusion is inescapable: no collective choice rule can satisfy certain minimal conditions of workableness and ethical acceptability.

10.8   The Borda Effect

It was pointed out in Section 10.6 that Condorcet's paradox of cyclic majorities cannot occur if the profile of individual preferences is single-peaked, which is bound to be the case if the voters evaluate the alternatives according to the same single criterion. Unfortunately, however, single-peakedness is no guarantee of freedom from paradox; a paradox originally discovered by Borda (1781) can arise even in a perfectly single-peaked profile of preferences. I have called this relatively little known paradox the "Borda effect" (Colman, 1979b, 1980; Colman and Pountney, 1978). The simplest manifestation of the Borda effect can be illustrated most easily by extending the example of Leftwich, Middleton, and Rightsman used earlier in this chapter. If the committee which has to reach a collective decision contains seven members, and if two of them are leftists, two are moderates, and three are rightists, then the profile of individual preferences is as shown in Table 10.5.

Table 10.5   Seven committee members' preference rankings of Leftwich (L), Middleton (M), and Rightsman (R)

| | Committee members | | | | | | |
| | Leftist | | Moderate | | Rightist | | |
	1	2	3	4	5	6	7
Most preferred alternative	L	L	M	M	R	R	R
Second ranked alternative	M	M	L	L	M	M	M
Least preferred alternative	R	R	R	R	L	L	L

The profile of individual preferences shown in Table 10.5 is perfectly single-peaked since the committee members all evaluate the candidates according to the same single criterion of "left-wing-ness". But which candidate ought to be chosen by a democratic collective choice rule? In a simple plurality vote, Rightsman would be chosen (if voting is sincere) because three voters (5, 6, and 7) consider Rightsman best while Leftwich and Middleton are each considered best by only two of the voters. But if the method of majority decision were used, that is, if the committee members voted on each pair of candidates in turn, then Middleton would be chosen. A majority of the committee members (3, 4, 5, 6, and 7) prefer Middleton to Leftwich, a majority (1, 2, 3, and 4) prefer Middleton to Rightsman, and a majority (1, 2, 3, and 4) prefer Leftwich to Rightsman. The collective preference scale is complete and transitive: MLR. The paradoxical character of the Borda effect comes into sharp focus when the following facts are appreciated: (i) Rightsman would be chosen by simple plurality voting, assuming that the voters were sincere; (ii) a majority of the voters (the leftists and moderates) prefer both Leftwich and Middleton to Rightsman. The plurality winner is the *least* preferred alternative in another plausible sense: a majority of the voters consider this alternative to be the worst of the three.

We may distinguish between two forms of the Borda effect (Colman and Pountney, 1978). The *strong Borda effect* occurs when there is a unique plurality winner and all of the losing alternatives are preferred to it by a majority of the voters. The *weak Borda effect* occurs when there is a unique plurality winner and at least one of the losing alternatives is preferred to it by a majority of the voters. A minimum of three alternatives are required for the effect to occur in either the strong or weak form, and when there are exactly three alternatives, the smallest number of voters necessary to produce the effect is seven. In a group consisting of eight voters choosing among three alternatives the Borda effect is (curiously) impossible, but in larger groups the possibility always exists.

Colman and Pountney (1978) and Colman (1980) have calculated some probabilities of the Borda effect under certain assumptions about profiles of preferences. In the simplest case of seven voters choosing among three alternatives, for example, the probabilities of the strong and weak effects assuming an equiprobable culture are 0.018 and 0.126 respectively. Exact probabilities in equiprobable cultures with larger numbers of voters choosing among three alternatives have been calculated by Gillett (see Colman and Pountney, 1978, pp. 16–17). Some of these results are summarized in Table 10.6.

Table 10.6   The Borda effect: theoretical probabilities in equiprobable cultures with three alternatives

	Numbers of voters												
	7	*8*	*9*	*10*	*11*	*12*	*... 100*	*101*	*... 200*	*201*	*... 300*	*301*
Strong effect	0.018	0	0.012	0.003	0.006	0.002	... 0.017	0.025	... 0.020	0.028	... 0.023	0.029
Weak effect	0.126	0	0.132	0.050	0.106	0.056	... 0.202	0.247	... 0.228	0.269	... 0.245	0.276

The results in this table reveal several interesting features. First, the probabilities of both effects tend to rise as the number of voters increases. Second, the probability of the weak effect is considerably higher than that of the strong effect in groups of all sizes. Third, both effects are more probable when the number of voters is odd than when it is even, and this remains true even in very large groups.

Monte Carlo simulations have been used by Fishburn (1974a) and Fishburn and Gehrlein (1976) in order to estimate the effect of varying the number of alternatives on the probabilities of the Borda effect (which they do not refer to by this name). Their results show that the probabilities tend to rise quite rapidly as the number of alternatives increases. They have also estimated probabilities under various other voting procedures (apart from plurality voting) of failing to choose an alternative when it is preferred to each of the others by a majority.

I have attempted elsewhere (Colman, 1980) to outline some of the properties of preference profiles that may be expected to maximize or minimize the likelihood of the Borda effect. The property of single-peakedness is not critical in this case, as the example of Leftwich *et al.* has demonstrated. It turns out, however, that any tendency towards uniformity of opinion among the voters reduces the probability below that expected in an equiprobable culture. If, for example, only one or two preference scales are represented among the voters, even the weak effect cannot occur. On the other hand, the probabilities of the weak effect are maximized in "completely discordant" cultures in which cyclic majorities are possible. In a seven-member committee choosing among three alternatives, A, B, and C, for example, the probability of a Borda effect is 0.126 in an equiprobable culture. If, however, the only preference rankings chosen by the voters are ABC, BCA, and CAB, then the probability is 0.288. The same figure applies to the other completely discordant culture characterized by the preference rankings ACB, BAC, and CBA. These are, of course, the preference rankings that produce cyclic majorities when they are evenly (or approximately evenly) distributed among the voters. The presence of a cyclic majority is in fact a sufficient condition, though not a necessary one, for the occurrence of the Borda effect.

The degrees of concordance or discordance which arise in real groups of voters are, of course, a matter for empirical investigation. In most attitudinal arenas a tendency towards concordance may be predicted. Similar socialization experiences, mass media effects, shared objective interests and the like all tend to promote a measure of uniformity of taste among individuals. Even in the area of purely aesthetic preferences, one may therefore hypothesize that the frequency of occurrence of the Borda effect will be less than the probability in an equiprobable culture. This hypothesis was tested in an empirical investigation under laboratory conditions (Colman, 1980). Seven-member groups of university students were invited to rank triplets of female first names, such as "Adeline, Agnes, Alice" according to liking. The hypothesis was confirmed: the observed relative frequency of the Borda effect was 0.020, which is significantly less than the theoretical probability premised on the assumption of an equiprobable culture (0.126). The strong effect was not observed in any of the 200 group decisions analysed in this investigation.

Empirical evidence has also been presented by Colman and Pountney (1978) to show that the Borda effect probably occurred in a number of constituencies in the 1966 British General Election. The preference rankings of the voters with respect to the three main political parties were estimated from the results of a very large opinion survey in which potential voters were invited to place the three main parties in order of preference. The election results, coupled with the estimated profiles of preferences in the 261 constituencies contested by these three parties alone, enabled the occurrence or non-occurrence of Borda effects to be determined. No instances of the strong effect were found, but the weak effect occurred in 15 of the constituencies, and in all but one case it was a Conservative Party candidate who was elected in spite of one of the others evidently being preferred to him or her by a majority of the voters. In the constituency of Chippenham, for example the Conservative Party candidate was elected with 18,275 votes; the Liberal Party candidate received 17,581 votes, and the Labour Party candidate received 10,257 votes. But the estimated percentage of voters in this constituency who preferred Liberal to Conservative was 54.62. More than 15 occurrences of the Borda effect would have occurred if all preference rankings had been equiprobable among the voters; there is evidently a tendency towards concordance in this attitudinal arena, as might have been expected. It is nonetheless disturbing that 15 candidates who failed to be elected to Parliament were probably preferred to the successful candidates by a majority of the voters.

10.9 Summary

At the beginning of this chapter a particular class of multi-person games governed by formal collective choice rules was identified. The primitive terms and axioms of collective choice theory were outlined in Sections 10.2 and 10.3, and the remainder of the chapter focused on various consequences which flow from the assumptions that some form of voting procedure is used and that the voters are sincere in so far as they always choose their maximax strategies. Three common voting procedures were examined in detail in Section 10.4; successive, amendment, and plurality procedures were shown to lead to different collective choices from the same sets of alternatives and individual preferences in some cases, and different voting orders were shown to affect the outcomes as well. Some implications for procedural "voting on voting procedures" were briefly sketched. In Sections 10.5 and 10.6 Condorcet's paradox of cyclic majorities was discussed in detail, and in Section 10.7 Arrow's proof that the paradox rules out the possibility of constructing a workable and fair voting procedure was outlined. Section 10.8 centred on a less well known but equally important paradox, the Borda effect, and some of its implications were discussed.

11

Strategic Voting

11.1 Optimal Voting Strategies

IT WAS pointed out in Chapter 10 that collective decisions governed by formal voting procedures may be viewed as N-person games. The players are the voters or voting blocs, the outcomes among which the players are assumed to have preferences are the alternatives which might be collectively chosen, and the rules are determined by the voting procedures. Chapter 10 was devoted chiefly to an examination of the properties of voting procedures, and the question of the strategies which the voters might choose was pushed into the background; in fact the assumption was made that the voters are sincere, that is to say that they always vote for the alternatives which they most prefer. Sincere voting does not, of course, ensure that one's most preferred alternative will be chosen; it merely offers the possibility that this alternative *might* be chosen *if* the other voters vote in certain ways. In other words, sincere voting corresponds to the choice of a maximax strategy.

Since maximax strategies are often quite irrational from any point of view, it is not unreasonable to assume that voters sometimes adopt insincere strategies. This assumption leads naturally to two interesting problems, one empirical and the other theoretical. The empirical problem is this: under what circumstances and in what ways do voters behave insincerely? The theoretical problem is: how far is it possible to go in defining rationality—that is, in determining optimal strategies—in voting games? The empirical problem has not been tackled by controlled investigations; the evidence to date (see Section 11.9) is entirely anecdotal. On the theoretical front, however, impressive progress has been made in recent years.

After the brief historical outline that follows, Sections 11.3, 11.4, and 11.5 are devoted to Farquharson's method of solving voting games, and more efficient methods are outlined in Section 11.6. Theoretical conclusions about strategic voting are summarized in Sections 11.7 and 11.8, empirical evidence is discussed in Section 11.9, and the chapter is summarized in Section 11.10.

11.2   Historical Background

Strategic voting is probably as old as democracy itself: it was apparently quite common in ancient Greece and Rome (Stavely, 1972). But no serious attempt was made to study its implications until the mid-1950s when Farquharson produced an analysis of the problem in the framework of game theory. The full publication of this work was delayed for many years,* but when *Theory of Voting* finally appeared in 1969, it immediately became a classic and inspired a number of further investigations of strategic voting. Although Farquharson's slim volume is written in deceptively simple language, it is extraordinarily difficult to understand, and it is marred by numerous typographical and logical errors (Niemi, in press, has listed 28 of these).

The remainder of this chapter is devoted to a clarification of the fundamental ideas behind Farquharson's approach, a brief examination of certain improvements that have been suggested, an outline of the main conclusions that have been reached regarding strategic voting, and a summary of empirical evidence.

11.3   Insincere Voting and Equilibrium Points

Let us return, for the sake of continuity, to the example of Leftwich, Middleton, and Rightsman outlined in Section 10.2. The political views of these three candidates, it will be remembered, are left-wing, moderate, and extremely right-wing respectively. A committee or an electorate has to choose one of them to fill a vacant political office. There are three distinguishable voters or voting blocs: the leftist voters' preference scale is LMR, the moderates' preference scale is MLR, and the preference scale of the rightists is RML. The voting blocs are of approximately equal size; in particular, no single bloc contains more than half the voters, so any two blocs combined can outvote the third. The leftist voting bloc is assumed to be the largest of the three, or if the largest blocs are of equal size, one of the leftists has an additional casting vote.

Three common voting procedures—successive, amendment, and simple plurality voting—were examined in Section 10.4. It was shown that, if voting is sincere, the amendment procedure inevitably results in the choice of Middleton, the plurality procedure leads inexorably to the choice of Leftwich, and the successive procedure can result in either Leftwich or

---

* Robin Farquharson was a South African with radical political views, and it is a matter of public record (Farquharson, 1968) that he was also a manic-depressive psychotic. He was arrested on one occasion for giving away £5 notes in a public house in Thame, and on another for stripping naked on the platform of the St John's Wood underground station in London (pp. 11, 67–8). By the time his epoch-making work on strategic voting appeared in print, he was living as a tramp in London, and he met a tragic and violent death shortly thereafter.

Middleton being chosen depending on the order in which the votes are taken. Since the rightists fail to obtain the outcome they consider best under any of the voting procedures, the question naturally arises: can they obtain a preferable result by voting insincerely, that is by adopting some non-maximax voting strategy? The same question may be asked about the leftists and moderates under those procedures that lead to non-optimal outcomes from their own points of view. If the answer is Yes, then the sincere outcomes are not equilibrium points in the game theory sense, since there exist players who can obtain preferable outcomes by unilaterally switching to insincere strategies.

Let us first examine successive voting and find out whether the sincere outcome is an equilibrium point. Suppose Middleton were singled out in the first ballot. The sincere outcome is then as follows:

First ballot: In favour of M, moderates;
                  Against M, leftists and rightists.
                  Result: L or R. Therefore
Second ballot: In favour of L, leftists and moderates;
                  In favour of R, rightists.
                  Final Result: L.

This outcome is the one that the rightists consider worst. It is obvious, however, that if the rightists had voted in favour rather than against Middleton on the first ballot while the others voted sincerely, the final result would have been Middleton after just one ballot. This would be a definite improvement from the rightists' point of view. In other words, the sincere outcome, L, is not an equilibrium point in the game: at least one of the players—the rightists considered as a single voting bloc—would have cause to regret the choice of a sincere strategy once the choices of the others were revealed. To put it differently, the sincere outcome is *vulnerable* to the rightists acting concertedly as a single player. It is easy to see, furthermore, that the sincere outcome is not vulnerable to either the leftists or the moderates, and that if the rightists were to vote strategically in the way indicated, the new outcome would not be vulnerable to any of the players. This means that if the rightists voted insincerely for Middleton on the first ballot, and if the others voted sincerely, the outcome would be an equilibrium point in the game, and none of the voters would have cause to regret the way they voted. Thus, by assuming that voting occurs in blocs, we have succeeded in identifying an equilibrium point which has a claim to being the rational outcome of successive voting.

Let us examine now an example of the amendment procedure. Suppose the substantive motion is "That Rightsman be appointed" and the amendment "That the word 'Rightsman' be replaced by 'Middleton'". Assuming that one of the three candidates must be appointed, the default alternative is Leftwich. The sincere outcome is:

First ballot: In favour of *M*, leftists and moderates;
                 In favour of *R*, rightists.
                 Result: *M*.
Second ballot: In favour of *M*, moderates and rightists;
                 Against *M* (i.e. in favour of *L*), leftists.
                 Final Result: *M*.

Is this sincere result an equilibrium point? The answer is No, because the leftists would have cause to regret their choice of strategy. Assuming again that voting occurs in blocs, i.e. that each bloc can be viewed as a single player, the leftists can obtain a preferable outcome by unilaterally switching to an insincere strategy. In the first ballot they are required to vote either for their second choice (*M*) or for the their third (*R*). If they are sincere they are bound to vote for *M*; but suppose they voted for *R* instead of *M* on the first ballot while the others stuck to their sincere strategies. In that case *R* would win the first ballot—so the amendment would be defeated—and if all voters were sincere on the second ballot between *R* and *L*, *L* would win. This outcome is preferred by the leftists to the sincere outcome: it is their favourite rather than their second choice. The sincere outcome is thus evidently not an equilibrium point since it is vulnerable to the leftists. But in this case neither is the new outcome when only the leftists vote strategically, for the rightists would then have cause to regret not voting for *M* rather than *R* on the first ballot, in which case (with the strategies of the other voters unchanged) the final result would be *M* rather than *L*, which they prefer. The outcome with both the leftists and the rightists choosing the prescribed insincere strategies is an equilibrium point upon which none of the players could unilaterally improve.

Under the plurality procedure, the sincere outcome is obviously Leftwich, because we have assumed that the leftists command more votes than either of the other blocs. But it is equally obvious that this outcome is vulnerable to the rightists and is therefore not an equilibrium point. To be specific, the rightists can vote en bloc in favour of Middleton instead of Rightsman; if the others vote sincerely this results in the choice of Middleton (the rightists' second choice) rather than Leftwich (their least favourite candidate). This strategic outcome is clearly an equilibrium point.

I have shown informally how equilibrium points can be found in voting games by assuming that voting always occurs in blocs. Two comments need to be made about equilibrium points, however. The first concerns voting games which satisfy the condition of complete information. If all of the voters are fully aware of the preferences of the others, then a non-equilibrium outcome can never occur provided that the voters choose their strategies rationally. For a non-equilibrium outcome is vulnerable to one or more players, and these players, if they are rational, will always take advantage of the opportunities for improvement which it presents. Secondly, all voting games possess equilibrium points in pure strategies. It is clear, for example, that if all voters choose the same alternative on each ballot—for example if all vote for L under the plurality procedure—then the result must be an equilibrium point since no voting bloc has the power to achieve a different outcome by unilaterally changing its strategy.

Unfortunately, the usual problems with equilibrium points as solutions to N-person games apply with equal force to the special case of voting games, in spite of the fact that pure strategy equilibria are known to exist. One difficulty arises from the fact that there are no known algorithms for finding all equilibrium points in an efficient way, that is apart from listing the entire set of outcomes and testing each one for stability. Another is that voting games, like other N-person games, often possess a number of equilibria, some of which seem more "rational" than others.

Consider the relatively simple example of three alternatives and three voting blocs choosing via plurality voting once again. Each player possesses three pure strategies: L (vote for Leftwich), M (vote for Middleton), and R (vote for Rightsman). Even assuming that there are only three players, in other words that the voters choose their strategies en bloc, there are no fewer than 27 ways in which the pure strategies of the leftists, moderates, and rightists respectively, may be combined: (L,L,L), (L,L,M), (L,M,L) and so on down to (R,R,R). Some of these combinations result in equilibrium outcomes, and others do not. The sincere outcome (L,M,R) is not, as we have seen, an equilibrium. If the rightists vote insincerely for M, on the other hand, the outcome (L,M,M) is an equilibrium. But this property alone is not compelling as a criterion for a rational solution, because there are other equilibria. For example, (L,L,L), (M,M,M), and (R,R,R) are also equilibrium points in the game because no single voting bloc can outvote the other two combined, thus none could rationally regret its choice of strategy if the outcomes were these. Some of the equilibria are obviously absurd. The combination (R,R,R), for example, could never occur if the voters were all rational: the leftists would never vote for Rightsman because voting for Leftwich or Middleton produces outcomes which in every logically possible contingency are at least as good (in their opinion) and sometimes better than those produced by voting for Rightsman. A rational solution to a voting game must clearly correspond to an equilibrium point, but we need some rational criteria for eliminating undesirable equilibrium points and arriving (if possible) at determinate solutions. In other words we need a method of finding optimal equilibrium points.

11.4   The Classical Solution: Dominance and Admissibility

The classical solution to voting games put forward by Farquharson (1969) treats each collective choice as a single N-person game, with the number of players equal to the number of distinct voting blocs. Only ordinal preferences among the possible outcomes are assumed, therefore mixed strategies cannot be evaluated: each player is assumed to choose a pure strategy. A pure strategy in the case of successive voting above, for example, is a choice from the following set:

(i) Vote for M on the first ballot; if M loses vote for R on the second ballot.
(ii) Vote for M on the first ballot; if M loses vote for L on the second ballot.
(iii) Vote against M on the first ballot; if M loses vote for R on the second ballot.
(iv) Vote against M on the first ballot; if M loses vote for L on the second ballot.

These four pure strategies exhaust the possibilities, and each provides a complete prescription for voting. There are also four pure strategies available to each voter in the amendment procedure (given three alternatives), and in the plurality procedure each voter obviously has just three pure strategies, which I have labelled L, M, and R.

The classical solution is arrived at by examining the entire game from the perspective of each player in turn and eliminating dominated strategies. A reduced version of the game from which the dominated strategies and their corresponding outcomes have been eliminated is then examined to see whether any strategies which were undominated in the original game are dominated in the first reduction. If so, they are deleted and the analysis proceeds to a second reduction. The process of eliminating dominated strategies continues until further reduction becomes impossible. Farquharson showed that this method leads to determinate solutions, in which only one ultimately admissible strategy survives for each player, in a surprisingly large percentage of cases. It is closely analogous to the method of solving certain two-person zero-sum games with saddle-points by successively eliminating dominated strategies; the method is insufficient for solving most two-person zero-sum games but adequate for most voting games. Farquharson's method is extremely cumbersome, but it is worth seeing how it works in the simplest case of the plurality procedure. This will incidentally justify the strategic solution which was arrived at informally above.

Table 11.1 shows the plurality procedure from the viewpoint of the leftists. Each of the 27 possible combinations of pure strategies is given in the table together with its corresponding outcome. The leftists choose one of the rows: they vote for L, M, or R. The choices of the other voting blocs deter-

Table 11.1   Plurality voting outcomes from leftists' viewpoint (leftists are most numerous)

Leftists vote for	Moderates/rightists vote for								
	L/L	L/M	L/R^s	M^s/L	M^s/M	M^s/R^s	R/L	R/M	R/R^s
*L^s	L^e	L	L	L	M^e	L^s	L	L	R
M	L	M	M	M	M^e	M	M	M	R
R	L	R	R	R	M	R	R	R	R^e

^{s} Sincere strategies or outcome.
^{e} Equilibrium points.
* Primarily admissible strategy for leftists.

mine the column: if the moderates vote for L and the rightists vote for R, for example, the final outcome is in column L/R. The body of the table thus shows the final outcome given all possible combinations of choices by the three voting blocs. The sincere strategies and outcome are indicated by footnotes; and the equilibrium points are also shown.

The equilibrium points in Table 11.1 offer an *embarras de choix*. The L^e in the top left-hand corner of the table, for example, denotes the outcome in which the leftists vote sincerely for L, the moderates vote insincerely for L, and the rightists vote insincerely for L: the final outcome is obviously L. Should this occur, none of the voting blocs would have cause to regret its choice of strategy since none could achieve a preferable outcome by unilaterally switching to a different strategy. But no fewer than four of the outcomes are equilibrium points, and they offer little guidance to the voters for the following reason: each voting bloc has at least one equilibrium corresponding to each of its pure strategies.

It is possible to show from Table 11.1, however, that the leftists have only one admissible strategy. Let us compare first of all the top two rows of the table, which correspond to the first two strategies of the leftists, bearing in mind that their preference scale is LMR. It turns out that in every contingency, that is in every column corresponding to a logically possible combination of strategies of the other players, the outcome is either the same or better from the leftists' viewpoint if they choose L rather than M. If the moderates and rightists both choose L (column L/L), for example, the outcome is the same whether the leftists choose L or M. If the moderates choose L and the rightists choose M (column L/M), then the leftists do better by choosing L than by choosing M because L wins instead of M and they prefer L to M. In every column, the leftists achieve an equal or preferable outcome by choosing the first row rather than the second. In other words, the leftists' strategy L dominates their strategy M.

Comparing each pair of rows in Table 11.1 in this way, we can quickly establish that the leftists' strategy M is dominated by L, and R is dominated by both L and M; the only undominated strategy available to the leftists is L.

An undominated strategy such as this is described as *primarily admissible* (the point of the qualifying adverb will emerge shortly). It is obvious that the leftists, if they are rational, must choose this primarily admissible strategy. To choose one of the dominated strategies would invite the possibility of an outcome possibly less desirable, and in no contingency preferable, to the outcomes resulting from the choice of their admissible strategy. The leftists are therefore well advised to exclude the dominated strategies M and R from consideration in deciding how to vote under the given plurality system. It is worth noting that they can reach this conclusion without any knowledge of the preferences of the other voters; it emerges from examining all logically possible outcomes from their own point of view alone.

In a sense, we have solved this game with respect to the leftists. Since they have only one primarily admissible strategy, it is their only rational choice. If a player possesses only one primarily admissible strategy in any game, then it is necessarily the case that this strategy dominates each of the others. In the classical theory of strategic voting, a dominant strategy of this kind is called a *straightforward* strategy, and a voting procedure which offers a voter a straightforward strategy is said to be a straightforward procedure for that voter. A straightforward strategy, being dominant, is a "sure thing" (uncon-ditionally best) strategy in game theory terms. It does not guarantee to a voter that his favourite alternative will be chosen, but it ensures that a failure to choose this alternative will not be the result of any strategic error on the part of this voter. But, as we shall see, a primarily admissible strategy is not necessarily straightforward: it is possible for a strategy to be undominated without itself dominating all of the others, though this is possible only if there are other undominated strategies as well. It is easy to prove (see Brams, 1975, pp. 69–70) that a voter under any voting procedure always possesses at least one primarily admissible strategy, namely sincere voting. If the voter has only one primarily admissible strategy, in other words if it is a straightforward strategy, then it must therefore be sincere. But a primarily admissible strategy may not be unique and may therefore not be straight-forward. This is illustrated in Table 11.2, in which we examine the plurality procedure from the viewpoint of the moderates.

Bearing in mind that the moderates' preference scale is MLR, we can search in Table 11.2 for their admissible strategies as before. A comparison of row L with row M reveals that they result in identical outcomes in some contingencies (e.g. column L/L), row M is preferable in some (e.g. column L/M), and row L is preferable in one contingency (column R/L). Clearly neither strategy dominates the other. A comparison of each pair of rows reveals that strategies L and M are primarily admissible: neither is domi-nated by any other strategy. Strategy R is not primarily admissible, since it is dominated by M (though not by L). The moderates have two primarily admissible strategies, L and M. Neither, however, is straightforward, since

Table 11.2   Plurality voting outcomes from moderates' viewpoint
(leftists are most numerous)

Moderates vote for	Leftists/rightists vote for								
	L^s/L	L^s/M	L^s/R^s	M/L	M/M	M/R^s	R/L	R/M	R/R^s
*L	L^e	L	L	L	M	M	L	R	R
*M^s	L	M^e	L^s	M	M^e	M	R	M	R
R	L	L	R	M	M	R	R	R	R^e

^{s} Sincere strategies or outcome.
^{e} Equilibrium points.
* Primarily admissible strategies for moderates.

neither dominates *all* of the others. One of the primarily admissible strategies, L, does not in fact dominate *any* of the others, although it is not dominated by any either. It is impossible at this stage to reach a decisive conclusion about which strategy the moderates will choose if they are rational. We may reasonably conclude, however, that they will not choose their primarily inadmissible strategy: they will not vote for R because the outcomes that result from doing so are in no cases better and sometimes less desirable from their point of view than those that result from their choice of the strategy M. We shall therefore delete the moderates' strategy R from the game; but first we must examine the situation from the viewpoint of the rightists (see Table 11.3).

The rightists' preference scale, it will be recalled, is RML. With this information, an analysis of Table 11.3 reveals that they have two primarily admissible strategies: voting for M and voting for R. Like the moderates, they have no straightforward strategy. If they are rational, however, they will not vote for L because they achieve outcomes at least as preferable, and sometimes more so, by voting for R: R dominates L.

Through arguments based on the principle of dominance, we have succeeded in finding a unique rational strategy for the leftists and in ruling out certain inadmissible strategies available to the other voting blocs or players. This has been achieved without any assumptions about the players' knowledge of the preference scales of the others. It is obviously not correct to say, as Brams (1975, 1976) does, that "if the voters have no information on each other's preference scales, they would have *no* basis on which to . . . formulate strategies" (1975, p. 59, my emphasis). But the solution is not yet complete, and in order to put the final touches to it we do need to assume complete information.

11.5   Sophisticated Voting

If each player has complete information about the preferences of the others, then each is in a position to examine a reduced version of the game con-

Table 11.3 Plurality voting outcomes from rightists' viewpoint
(leftists are most numerous)

Rightists vote for	Leftists/moderates vote for								
	L^s/L	L^s/M^s	L^s/R	M/L	M/M^s	M/R	R/L	R/M^s	R/R
L	L^e	L	L	L	M	M	L	R	R
$*M$	L	M^e	L	M	M^e	M	R	M	R
$*R$	L	L^s	R	M	M	R	R	R	R^e

^{s} Sincere strategies or outcome.
^{e} Equilibrium points.
$*$ Primarily admissible strategies for rightists.

structed by assuming that only primarily admissible strategies will be chosen. This reflects an assumption on the part of each player about the likely behaviour of the others. But the assumption that the others will avoid using dominated strategies seems reasonable enough: if a player has sufficient common sense to avoid the strategies that can only be to his disadvantage, then he ought to credit the other players with equal good sense.

It is possible to search for admissible strategies in the reduced version of the game from which all primarily inadmissible strategies have been deleted. A strategy that is admissible for a player in this first reduction of the game is called, for obvious reasons, a *secondarily admissible* strategy. A secondarily admissible strategy is premised on the assumption that all players will use only primarily admissible strategies. The analysis can be repeated, if necessary, by deleting secondarily inadmissible strategies from the first reduction of the game, and searching in the resulting second reduction for 3-arily admissible strategies. Farquharson's classical approach proceeds in this way, searching for m-arily admissible strategies for each player on the assumption that the others will choose only $(m - 1)$-arily admissible strategies, until a determinate solution is found, if such a solution exists. The type of voting behaviour which this implies is known as *sophisticated voting*. In the simple plurality procedure which we are analysing, a determinate solution emerges from the first reduction, which is shown in Table 11.4.

No choice of strategies for the leftists is shown in Table 11.4, since they have no choice after their primarily inadmissible strategies have been deleted: it is assumed that they choose their straightforward strategy L. The game has been further reduced by deleting the primarily inadmissible strategies of the moderates and rightists. On the assumption that only the primarily admissible strategies indicated will be chosen, it is clear that the game can be completely solved. From the moderates' viewpoint, row M represents the only undominated—and hence secondarily admissible—strategy. The rightists, for their part, have only one undominated column: M. The secondarily admissible strategies of these players are in fact ulti-

Table 11.4   Plurality procedure after deleting primarily inadmissible strategies (leftists are most numerous)

Moderates (MLR) vote for	Rightists (RML) vote for	
	$**M$	R^s
L	L	L
$**M^s$	M^e	L^s

^{s} Sincere strategies or outcome.
^{e} Equilibrium point.
** Secondarily admissible strategies.

mately admissible, since further reduction of the game is unnecessary. The straightforward strategy of the leftists, together with the secondarily admissible strategies of the moderates and rightists, converge on a unique equilibrium point *LMM*. Since this sophisticated solution is an equilibrium point—as it is always bound to be—none of the players has both the will and the power to avoid it, and none has reason to regret its choice of strategy.

We may therefore conclude that if the members of the committee or electorate are rational or sophisticated, then the leftists will vote straightforwardly for Leftwich, and the moderates and rightists will vote for Middleton. This means that Middleton will be elected to the vacant post in spite of the fact that the leftists command the largest bloc of votes, or that the additional casting vote belongs to a leftist if the blocs are of equal size. If voting were sincere, of course, Leftwich would be elected; the sophisticated outcome, even in this extremely simple example, is rather unexpected.

11.6   Anticipated Decisions and Multistage Solutions

Farquharson's classical method of solution can in principle be applied to any voting game; but it has two major drawbacks. First, it is extremely cumbersome to apply. Quite an elaborate analysis was required to solve even the simple game above. Other voting procedures, such as successive and amendment voting, are much more complicated to analyse even in the three-alternative case, since each voter is confronted with four rather than three pure strategies in the original games. When there are more than three alternatives the classical method of solution under sequential voting procedures becomes unmanageable without the use of a computer. Secondly, the classical method fails to produce a determinate solution in a significant minority of cases under the plurality procedure: there comes a point beyond which the game cannot be reduced by eliminating dominated strategies. If the preference scales of the three voting blocs are *LMR*, *MRL*, and *RML*, for example, Farquharson's method fails to produce a determinate solution.

It never fails in this way, however, with successive or amendment procedures.

These problems have been partly overcome by the discovery of more efficient methods of solving voting games. The difficulties that arise in dealing with complicated voting procedures involving a large number of alternatives, voting blocs, and ballots can be circumvented by multistage game analyses, in which the entire voting game is broken down into a number of smaller games which are solved very easily, and methods have been devised for producing determinate solutions in some (though not all) plurality voting games which cannot be solved by classical methods. I shall outline each of these contributions separately.

Miller (1977) made some headway towards simplifying the analysis of complicated voting games via graph theory. McKelvey and Niemi (1978) and Bjurulf and Niemi (1981) have gone further in providing an extremely simple and efficient method of analysis which works for all binary voting procedures, no matter how complicated. Binary voting procedures are those like successive and amendment voting in which each voter has to choose between two alternatives on each ballot. Apart from plurality voting, most procedures in common use are binary in this sense. McKelvey and Niemi's solution is based upon a representation of the voting game as a hierarchy of minor games, which can be depicted in the form of a decision tree. Typical decision trees for the successive and amendment voting procedures are shown in Figure 11.1.

The terminal nodes in Figure 11.1 are labelled with the appropriate collective choices. The non-terminal nodes are labelled G_1, G_2, and (in the amendment procedure) G_3: these nodes represent particular ballots and they are treated as minor games embedded in a larger voting game by McKelvey and

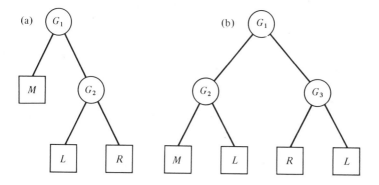

FIG. 11.1 Decision trees representing (a) the successive procedure with M presented in the first ballot, and (b) the amendment procedure with the first ballot between M and R. Non-terminal nodes G_1, G_2, and G_3 correspond to minor games.

Niemi, and by Bjurulf and Niemi. In the successive procedure (Figure 11.1a), for example, node G_1 corresponds to the first ballot in favour or against M; if a majority favour M the corresponding terminal node is reached, otherwise the outcome of the first minor game G_1 is a second, G_2, which represents a ballot between L and R. In the amendment procedure (Figure 11.1b), the first minor game G_1 is always followed by a second minor game, either G_2 or G_3 depending on the outcome of G_1.

Consider the minor game G_1 under the successive procedure. The outcomes of this game are shown in Table 11.5. This minor game could obviously equally well be represented from the moderates' or rightists' viewpoints; this simply entails a rearrangement of the labels. It is obvious from this representation of G_1 that if the leftists prefer M to G_2, then they have only one rational (dominant) strategy, namely voting for M; conversely if they prefer G_2 to M they would be rational if they voted against M. But G_2 is a game, not an outcome. In order to determine their preference of M vis-à-vis G_2 the players have first to solve G_2; they can then substitute this "anticipated decision" (Miller, 1977) or "sophisticated equivalent" (McKelvey and Niemi, 1978) in the payoff matrix of G_1.

Table 11.5 The minor game G_1 under the successive procedure: voting for or against M from the leftists' viewpoint

Moderates	For		Against	
Rightists	For	Against	For	Against
Leftists				
For	M	M	M	G_2
Against	M	G_2	G_2	G_2

The multistage solution therefore begins at the bottom of the decision tree and works upwards by stages. Minor games at the lowest level, whose outcomes are all determinate collective choices and not further minor games, can be solved at a glance from a knowledge of the voters' preferences. Whichever of the two outcomes is preferred by a majority of the voters—or by the voter with the additional casting vote if opinions are split evenly—is bound to be chosen at this point; there is no scope for strategic voting on such a ballot because there are only two alternatives. Under the successive procedure at G_2, for example, L is bound to be chosen because a majority of the voters (the leftists and moderates) prefer L to R. Since the minor game G_2 thus has the inevitable outcome L, this anticipated decision can be substituted for G_2 in Table 11.5 (which represents G_1), and it is immediately obvious that the leftists will vote against M if they are rational in view of the fact that this is a straightforward strategy.

The computation of the anticipated decisions at each node is extremely easy, and it is not necessary to set out the corresponding payoff matrices as in

Table 11.5. Those at the lowest level are found by determining which of the corresponding pair of outcomes is preferred to the other by a majority of the voters (taking into account casting votes if necessary). These anticipated decisions are then placed in their correct positions at the next level up the decision tree and the whole process is repeated for the minor games immediately above them, and so on. However complicated the decision tree may be, and however many voting blocs there are, this method of solution rapidly produces an anticipated decision for each non-terminal node in the tree. In Figure 11.1a, for example, G_2 is replaced by L because a majority prefers L to R. Then G_1 can be replaced by M because a majority prefers M to L. In Figure 11,1b, G_2 becomes M and G_3 becomes L, then G_1 becomes M. Having filled in the nodes of the decision tree in this way, the optimal strategies for each player are immediately obvious: at each node a rational player will always vote for the outcome or anticipated decision that he or she prefers.

The multistage solution of binary voting games is vastly simpler than the classical method, and it has the added virtue of reflecting more realistically the way rational voters probably think. Strategic voters probably think in something like the following manner: "Should I vote for or against M? Well, what is likely to happen if I vote against M and all the other voters pursue their own interests rationally? Clearly L will win on the second ballot, so I'm better off voting against M on the first". In this hypothetical soliloquy, "what is likely to happen" corresponds to the multistage notion of the "anticipated decision".

An interesting question is this: does the multistage solution always correspond to the classical solution? The answer is rather unexpected: the two methods always yield the same determinate outcomes, but the optimal strategies which they prescribe for the players sometimes differ. Neither method ever fails to produce a complete solution to any binary voting game, however. The multistage method is easy to apply under complicated binary procedures with numerous alternatives and voting blocs (players), provided only that the preferences of the voters are known. The existence of cyclic majorities (see Section 10.5) presents no problem since it is still easy to determine which of any pair of alternatives will win on any ballot; the anticipated outcomes and the optimal strategies can then be determined in the usual way.

The theory of strategic voting under the plurality procedure is not in quite such a healthy state, and I shall comment on it only briefly. Apart from Farquharson (1969), Joyce (1976), and Niemi and Frank (1980), no-one appears to have ventured into these theoretical swamps. The problem with the classical solution in this case is not so much one of cumbersomeness, but the fact that determinate solutions fail to emerge in certain cases. The Niemi–Frank approach represents a slight improvement in so far as it yields solutions in all situations in which the classical method does so, and it succeeds in a proportion of cases in which the classical method fails. In situations in which both methods yield determinate outcomes, these outcomes are apparently identical, though this has not been proved. The new method can therefore be viewed as an extension of the classical method. Niemi and Frank have shown that their procedure is effective in the overwhelming majority of cases involving three alternatives, but it is not clear how it can be extended to voting games with a larger set of alternatives.

With three alternatives, the essential ideas can be sketched as follows. Each voting bloc (player) examines the sincere outcome and determines whether it can obtain a preferable outcome by ulilaterally switching to some insincere strategy. If not, the sincere outcome is in fact a sophisticated equilibrium. If one player can unilaterally improve on the sincere outcome by adopting an insincere strategy, then this insincere strategy is substituted in the game and the analysis reverts to the beginning. If two players can each unilaterally improve on the sincere outcome, then Niemi and Frank show that these two players may be locked in a Battle of the Sexes game, and the solution may therefore be indeterminate in pure strategies. This approach to strategic voting under the plurality procedure has certain advantages over the classical method, but it lacks the simplicity and intuitive persuasiveness of the multistage solution to binary procedures. Further work needs to be done in this area.

11.7   General Results on Strategic Voting

Farquharson (1969) examined the results of sincere and strategic voting among three alternatives and three approximately equal voting blocs under the successive, amendment, and plurality procedures (and a few less common ones). Under the successive procedure, he found that sincere and strategic voting lead to different collective choices in one-half of all possible cases (preference scales and voting orders). The order of voting always influences the final outcome under this procedure if voting is sincere: the alternative voted upon first never wins. If voting is strategic, however, the order of the successive ballots seldom makes any difference, and the alternative voted upon first wins in fully one-half of cases.

With three alternatives and three voting blocs under the amendment procedure, Farquharson showed that the sincere and strategic outcomes are different in one-quarter of cases. If voting is sincere, the order of balloting makes a difference to the final outcomes only when cyclic majorities (see Section 10.5) are present, and an alternative presented in the first ballot wins in three cases out of four. If voting is strategic, the order of balloting is critical only in the same special cases, and an alternative presented in the first ballot wins in five cases out of eight.

Under the plurality procedure, restricting himself again to three alternatives and three voting blocs, Farquharson found that the sincere and strategic outcomes differ in three cases out of four. The largest voting bloc— or the voter with an additional casting vote in the case of equal-sized blocs— always obtains its favourite outcome when voting is sincere. If strategic voting is used, however, the favourite alternative of the strongest voting bloc wins in only one case out of four, and in two further cases the classical method of solution produces an indecisive result: any of the alternatives might win.

Using more efficient methods of analysis, several researchers have extended Farquharson's preliminary results to cover more complicated cases. To begin with, Kramer (1972) has proved that the sincere and strategic outcomes are identical under any binary voting procedure such as successive or amendment voting if the preference scale of each voter is *separated* at each ballot. The idea of separability, first introduced by Farquharson, is extremely simple. In Figure 11.1b, for example, the first ballot G_1 is between M or L on the one hand and R or L on the other. This ballot separates a preference scale such as MLR because it can be divided into the appropriate pairs (M,L), (L,R) without disturbing the original order. The scale LMR, on the other hand, cannot be separated into the required sets without disturbing the preference order. Since this procedure fails to separate the leftists' preference scale on the first ballot, the sincere and strategic outcomes may not be identical according to Kramer's theorem, and they are indeed different as was shown earlier. In complicated binary voting procedures with numerous ballots, alternatives, and voting blocs, this result enables a rapid check to be made to discover whether strategic voting may lead to a different outcome from sincere voting.

Miller (1977), McKelvey and Niemi (1978), and Bjurulf and Niemi (1981) have extended Farquharson's preliminary conclusions about the effects of voting orders under binary procedures. Although the order of voting is frequently critical for sincere voting, McKelvey and Niemi proved that this is not so for strategic voting: in that case the same outcomes arise from all possible voting orders unless a cyclic majority exists in the profile of voters' preferences. In cases where one of the alternatives is preferred to each of the others by a majority of the voters, in other words where a cyclic majority does not exist, McKelvey and Niemi have shown that this alternative is invariably the final outcome of strategic voting, although sincere voting may lead to one of the other alternatives being chosen, or to what was called in Section 10.8 a Borda effect. A Borda effect never occurs under binary voting procedures if voting is strategic unless a cyclic majority exists; if voting is sincere it may occur, though not under the amendment procedure. In situations involving cyclic majorities, Bjurulf and Niemi have established that it is always possible to find a voting order under successive or amendment procedures which ensures that a particular alternative will win given sincere voting, and a voting order which ensures that it will win given strategic voting. They have also refined Farquharson's preliminary findings regarding the effect of the position of an alternative in the voting order on its likelihood of being chosen. It turns out to be not strictly true in general that an alternative voted on late under a binary procedure is *always* at an advantage if voting is sincere, and an alternative is not *always* at an advantage if it is voted on early by strategic voters. A rational committee member or politician would, however, be well advised to try to ensure that his favourite amendment is voted on last if voting is sincere, and if voting is strategic to avoid its being voted on last. Under the successive procedure similar advice can be given, but in this case it is important to try to have one's favourite alternative presented in the very first ballot if voting is strategic.

Turning now to the plurality procedure, a (smaller) number of general conclusions are justified. First, there is the paradox of the chairman's vote. If voting is sincere, then the favourite alternative of the largest voting bloc, or the chairman's favourite alternative in the event of a tie, is bound to win. It is therefore rational to seek extra power in sincere decision making bodies. If voting is strategic, however, the largest voting bloc or the chairman may be positively disadvantaged, as in the example discussed in Section 11.5. Depending on the profile of preferences among the voters, it may be prudent to shun extra power in a sophisticated decision making body (this is a somewhat weaker prescription than the unsound one given by Farquharson and criticized by Niemi, in press). There is often a tendency for strategic voters to "gang up" against the most powerful voter or voting bloc, thereby ensuring that the wishes of this voter or bloc are bound to be frustrated. This paradox provides a vivid and simple refutation of the dogma that the fittest always survive; a complicated refutation, also based on game theory, was given by Shubik (1954), who showed that the worst shot in a three-person duel has the best chance of survival.

Some further general conclusions about the plurality procedure relate to the Borda effect discussed in Section 10.8. If voting is sincere, then a strong or weak Borda effect can occur, given certain profiles of preferences, provided that there are seven voters or more than eight voters. If voting is strategic, on the other hand, Niemi and Frank (1980) have shown that a strong Borda effect cannot occur: the winning alternative may still be one to which *some* of the losing alternatives are preferred by a majority of the voters, but it can never be one to which *all* of the others are preferred by a majority. This is an important result, although it is not as striking as the corresponding theorem about binary procedures mentioned earlier, which shows that in those cases even a weak Borda effect is impossible if voting is strategic unless a cyclic majority exists.

11.8   Is Strategic Voting Unfair?

For at least two centuries the view seems to have prevailed that strategic voting is somehow deceitful, unfair, and contrary to the spirit of democratic collective decision making. Shortly before the outbreak of the French Revolution, Borda suggested a new voting procedure which was designed to avoid certain undesirable paradoxes. When it was pointed out to him that his new procedure was vulnerable to strategic voting, he exclaimed "My scheme is intended only for honest men!" (quoted e.g. in Kelly, 1978, p. 65). During Victorian times, C. L. Dodgson (Lewis Carroll) lamented the apparently common tendency among voters to adopt a "principle of voting which makes an election more of a game of skill than a real test of the wishes of the electors" (quoted by Farquharson, 1969, p. 17), and he thought it would be better that "elections be decided according to the wish of the majority than of those who happen to have most skill at the game" (*ibid.*). More recently, the well-known American political scientist William H. Riker (1961) expressed the view that "there may be nothing wrong with lying as a political strategy, but one would not, I assume, wish to give a systematic advantage to liars" (p. 905). In the light of what is now known about the consequences of strategic voting, it is worth pausing to consider whether "dishonest, skilful lying" or strategic voting deserves such a bad press, and whether honest, clumsy sincerity is really more desirable from an ethical or democratic standpoint.

Consider first the problem of the Borda effect. Sincere voting allows the obviously undesirable possibility that an alternative may be chosen under many voting procedures in spite of the fact that a majority of the voters prefer each of the losing alternatives to the winner. Strategic voting rules this possibility out and ensures that no such unsuitable alternative can ever become the collective choice under any voting procedure in common use. Under binary procedures, in fact, even a weak Borda effect is rendered impossible by strategic voting unless a cyclic majority exists: *none* of the losing alternatives can be preferred by a majority to the winner. If a cyclic majority exists among the voters, then a Borda effect is obviously inevitable, but even in that case Bjurulf and Niemi (1981) have shown that sophisticated voting leads to "better" collective choices, that is, those which are ranked higher in the collective preference scale than the alternatives chosen by sincere voting. These considerations surely argue strongly in favour of the ethical and democratic desirability of strategic voting.

A second powerful argument in favour of strategic voting centres on the problem of voting orders under binary procedures. The order in which the ballots are held ought not to determine the collective choices in a fair system, yet it very frequently does if the voters are sincere. This is extremely rare, however, if voting is strategic, and can happen only if a cyclic majority exists.

It was shown in Section 10.4 that procedural votes, in which the order of balloting is chosen democratically, can sometimes prevent the will of the majority from being thwarted by an unfavourable voting order, even if voting is sincere. But procedural votes of this kind are frequently disallowed, and socially undesirable alternatives are therefore chosen by sincere decision making bodies. Brams (1975, pp. 83–5) has, however, proved that an alternative which fails to be chosen by sincere voters, but which would have been chosen under a different order of voting favoured by the majority, will win under the original "unfavourable" voting order provided that strategic voting is used. In other words, the voters who are disadvantaged by a particular order of (sincere) voting do not need recourse to a procedural remedy; they can secure a satisfactory outcome under the given order of balloting by adopting sophisticated strategies. Thus if a procedural vote is not permitted, the resource of strategic voting ensures that the will of the majority prevails.

Those who disapprove of strategic voting and do not "wish to give a systematic advantage to liars" often attempt to devise new voting procedures which are claimed to be strategy-proof. Steen (1980) has, for example, argued in favour of "approval voting" on these grounds. (Under approval voting, each voter is invited to indicate as many alternatives as he considers acceptable; it is a case of "one man, n votes".) But theorems have been proved to the effect that no voting procedure can be strategy-proof (in the sense of offering the voters no incentive to vote insincerely) without violating some more fundamental condition of democratic acceptability. Gibbard (1973) and Satterthwaite (1973, 1975) were the first to obtain this important result; they showed independently that any strategy-proof voting procedure is necessarily dictatorial. Further theorems along the same lines have been proved by Zeckhauser (1973), Pattanaik (1976), Gärdenfors (1976), Barbera (1977), and Kelly (1978, chap. 6). Even if one has a strategic voting phobia, therefore, it is a waste of time trying to devise an acceptable strategy-proof voting procedure: the quest is as vain as that for a perpetual motion machine.

In a democracy, we have to learn to live with strategic voting whether we like it or not. It is by no means obvious why so many people evidently do not like it. Strategic voting does not merely serve selfish interests; I have shown that in many circumstances it protects the interests of the majority against selfish minorities. In most games that occur in everyday life it is taken for granted that rational decision makers are likely to choose the strategies that common sense dictates, and such behaviour if seldom considered devious or underhand. It seems strange that the irrational choice of maximax strategies, in other words sincere voting, should be expected of intelligent voters, or that they should be slandered for voting rationally. A military general who adopted maximax strategies by choosing actions solely according to the most favourable possible outcomes given certain logically possible actions on the part of the enemy, would justifiably be regarded not only as a danger to himself and others, but also as a fool.

11.9 Empirical Evidence

Empirical investigations of strategic voting under controlled conditions have not been performed, but a number of examples of what appears to have been strategic voting have been reported anecdotally. These examples depend upon assumptions regarding the voters' preference scales and inferences about their voting strategies. Only a selection of some of the most frequently quoted examples will be mentioned here.

The following example, originally cited by Farquharson (1969), is taken from the letters of Pliny the Younger and refers to an incident in the Roman Senate:

> The consul Afranius Dexter had been found slain, and it was uncertain whether he had died by his own hand or at those of his freedmen. When the matter came before the Senate, Pliny wished to acquit them; another senator moved that they should be banished to an island; and a third, that they should be put to death (Farquharson, 1969, pp. 6–7; Pliny's letter is reproduced in full, *op. cit.*, pp. 57–60).

There appear to have been three opinion blocs within the Senate. The acquitters' preference scale was (obviously) ABC, that is, acquit, banish, condemn to death. The banishers' scale was BAC, and that of the condemners CBA. Pliny insisted that the decision should be reached by simple plurality voting. The outcome of sincere voting would have been A, since the acquitters constituted the largest voting bloc. The decision actually reached, however, was B, which corresponds to the outcome of strategic voting. If the preference scales were the ones (very plausibly) assumed by Farquharson, then there is every reason to believe that the senators voted strategically. This example is formally identical to the one examined in Sections 11.3 to 11.5 above.

A number of more recent examples are taken from American elections. Riker and Ordeshook (1973, p. 98) and Brams (1975, chap. 2) have, for example, argued that sincere voting may have determined the outcome of the 1912 presidential election. Although voting was apparently sincere in this case, the preference scales that the authors postulate imply that Roosevelt rather than Woodrow Wilson would have been elected had the voters adopted strategic methods of voting. In the 1948 presidential election, on the other hand, strategic voting seems to have occurred. According to Downs (1957), "some voters who preferred the Progressive candidate to all others nevertheless voted for the Democratic candidate" because sincere voting "ironically increased the probability that the one they favoured least [the Republican candidate] would win" (p. 47). In the event, Truman (Democrat) narrowly defeated Dewey (Republican), and Wallace (Progressive) got only 2.4 per cent of the popular vote.

During the 1970 election for a senator in New York, Ottinger (Democrat), Goodell (Republican), and Buckley (Conservative) were the three candidates. The Goodell supporters, who were in a minority, would probably have preferred Ottinger rather than Buckley to emerge as the collective choice. Had they all voted strategically for their second choice, Ottinger, then he would probably have won the election. Some Goodell supporters did vote strategically in this way (Niemi and Riker, 1976, pp. 24–5), but most voted sincerely and Buckley was therefore elected.

What is most striking about the above examples from American politics is the apparent unwillingness of most American voters to adopt voting strategies that serve their own best interests. This is no doubt bound up with the unfavourable connotations of strategic voting referred to earlier. In French labour elections which are held under proportional representation, on the other hand, there is indirect evidence (Rosenthal, 1974) suggesting that inhibitions about rational voting behaviour are weaker, and strategic voting is quite common. Some further empirical examples, together with some of those that have been outlined in this section, are discussed entertainingly and in considerable depth by Brams (1975, 1976, 1978).

11.10   Summary

This chapter began by focusing on the problem of defining rational voting strategies in contrast to the sincere maximax strategies which were assumed in Chapter 10. After a brief account of the historical origins of strategic voting and its investigation in Section 11.2, the discussion turned in Section 11.3 to equilibrium points in voting games. The usual problems with equilibrium points in N-person games were shown to exist: there are usually several to choose from and some seem more "rational" than others. In Sections 11.4 and 11.5, Farquharson's classical solution, via successive elimination of dominated strategies to find a "sophisticated" outcome, was explained in detail. Two major difficulties with this method, cumbersomeness and a failure to find determinate solutions to some plurality votes, were mentioned at the start of Section 11.6; the rest of that section centred on more efficient methods. The multistage game analysis was shown to be especially simple and powerful for binary procedures. Section 11.7 was devoted to general results on strategic voting, and in Section 11.8 the claim that strategic voting is unfair was challenged in the light of these results. Strategic voting was shown to lead to more "democratic" and ethically acceptable collective choices than sincere voting. Finally, Section 11.9 contained an outline of some empirical (though purely anecdotal) examples of sincere and strategic voting in ancient and modern times.

12

Theory of Evolution: Strategic Aspects

12.1 Historical Background

THIS CHAPTER focuses on applications of game theory to problems of biological evolution. The basic ideas were set forth in an unpublished paper on ritualized fighting in animals by George Price and subsequently developed by John Maynard Smith (1972; Maynard Smith and Price, 1973), but they can be traced back through the work of W. D. Hamilton (1967) on the evolution of sex ratios to R. A. Fisher (1930). The theoretical principles and supporting empirical evidence have been reviewed several times (Maynard Smith, 1974, 1976a, 1978a, 1978b, 1979; Parker, 1978; Lazarus, 1982). Some commentators, including the influential biologist Richard Dawkins (1976), have expressed the view that this work may represent "one of the most important advances in evolutionary theory since Darwin" (Clutton-Brock and Harvey, 1978, p. 139).

Only a few of the simplest evolutionary games and a selection of the empirical evidence are discussed in this chapter since fuller accounts are readily accessible in the publications listed above. In the following section the fundamental concept of strategic evolution is introduced and the relevance of game theory to natural selection is explained. Section 12.3 is devoted to Maynard Smith's two-person models of animal conflicts and the notion of an evolutionarily stable strategy, and an improved multi-person game model is outlined in Section 12.4. Some relevant empirical evidence is discussed in Section 12.5 and the chapter is summarized in Section 12.6.

12.2 Strategic Evolution

In its modern neo-Darwinian form, the theory of evolution rests on two major premises: (a) that evolution consists of changes in the frequencies of genes in populations of plants and animals; and (b) that the frequency of a particular gene increases if it increases the *Darwinian fitness* of its possessors. The Darwinian fitness of a genotype is defined simply as its expected number of surviving offspring, which in turn determines the number of gene copies that it transmits to future generations.

Offspring tend to resemble their parents, but the mechanisms of genetics, which are now well understood, ensure that hereditary variations in anatomy, physiology, and behaviour are always present in a population. A given environment poses "problems" of adaptation that organisms need to "solve" in order to survive, and evolution through natural selection, according to the premises mentioned in the previous paragraph, ensures that the fittest genotypes increase at the expense of the relatively less fit.

The essential concepts of game theory can be placed in one-to-one correspondence with the elements of the theory of natural selection. The players correspond to the individual organisms—animals, plants, viruses and so on—in a population. Each organisms's strategy is determined by its genotype, and the payoffs are the resulting changes in the Darwinian fitness of the organisms. No assumption needs to be made that the players (organisms) "choose" their strategies (genotypes) deliberately or rationally, of course. But different strategies can result in different payoffs, and it is well known that natural selection can mimic deliberate choice. In view of the fact that Darwinian fitness is defined numerically and unambiguously, one of the major difficulties in game theory applications—the problem of assigning numbers to the payoffs—is conveniently avoided in evolutionary applications. A human decision maker's preferences among the outcomes of a game can often be quantified only roughly and with some difficulty, but an organism's expected number of surviving offspring is inherently quantitative.

Classical evolutionary models are usually non-strategic in character. The fitness of a genotype is traditionally assumed to depend solely upon its own intrinsic characteristics; the characteristics of other individuals in the population are assumed to be either fixed or irrelevant to the model. In the terminology of game theory, classical evolutionary models assume that an individual's expected payoffs are determined solely by its own strategy choices, and the corresponding games are therefore one-person games against Nature. Lewontin (1961) was the first to formulate evolutionary problems explicitly in terms of one-person game models (he suggested that species should, and do, adopt minimax strategies against Nature) but the same non-strategic assumptions are implicit in most of the literature on natural selection.

It is obvious, however, that the fitness of a genotype sometimes depends on the genetic composition of the population or, in the terminology of modern population genetics, it is *frequency-dependent*. This is typically the case with hereditary behavioural dispositions—the consequences of different types of courtship behaviour, for example, obviously depend on how others behave—but in principle it can be true also with regard to anatomical and physiological characteristics. Whenever the fitness of a particular genotype depends on the genetic composition of the population, we may translate into the language of game theory and say that the payoff resulting from a particular strategy depends on the strategies of the other players; and in such cases multi-person game models are clearly required. Natural selection poses genuinely strategic problems to organisms with regard to certain kinds of traits, but this fact was virtually ignored by evolutionary theorists for more than a century.

Strategic aspects of natural selection were first recognized in the evolution of sex ratios. The strategies available to individuals are simply the relative numbers of male and female offspring they may be genetically programmed to produce. Most animals and plants with distinct sexes have evolved to stable equilibria in which approximately equal numbers of males and females are produced. Once it is recognized that this requires explanation, the solution is not difficult to find. Fisher (1930) was the first to point out that the sex which is likely to have more offspring, and thus to transmit more gene copies to future generations, will be favoured by natural selection. A simplified version of Fisher's argument is as follows. In a population with a preponderance of females, an individual that produced only sons would end up with the most grandchildren, since all offspring need fathers, and genes predisposing parents to produce sons would therefore increase in frequency. In a population consisting mostly of males, on the other hand, an individual would maximize its payoff by producing only daughters. In either case, the sexual imbalance in the population would be corrected by natural selection. The population would therefore evolve to the stable equilibrium point at which equal numbers of males and females are produced and the production of sons and daughters yield the same payoffs.

This type of analysis was taken a step further by Hamilton (1967) who used the methods of game theory to explain the "extraordinary sex ratios" found in certain parasitic wasps and other insects. The females of this group of insects store sperm and can determine the sex of each egg they lay by fertilizing it or leaving it unfertilized; fertilized eggs develop into females and unfertilized eggs into males. Some parasitic wasps lay their eggs in the larvae of other insects such as caterpillars. The wasps mate immediately upon hatching, and after mating the males die and the females disperse.

If only one wasp lays her eggs in each larva, her optimal strategy is clearly to leave exactly one unfertilized egg in a cluster since one male can mate with all of the females when the eggs hatch. But if more than one female can lay her eggs in the same larva, then the optimal ratio of female to male eggs is an inherently strategic problem: it is essential to leave an unfertilized egg if none of the others are likely to do so but this strategy is not necessarily optimal otherwise. A large predominance of females is in fact found in these species, and Hamilton (1967) was able to show that the observed sex ratios are usually optimal, or very nearly so, from a game theory standpoint.

12.3 Animal Conflicts and Evolutionarily Stable Strategies

Many species of animals avoid escalated "no-holds-barred" fights with one another and engage instead in relatively harmless forms of conventional combat to settle conflicts over valuable resources such as food, mates, territory, or positions in dominance hierarchies. Bighorn rams leap at each other head-on, although more damage could be inflicted on an adversary by charging him in the flank. In many species of fish the combatants seize each other by the jaws, which are protected by leathery skin, rather than biting elsewhere where more injury would be inflicted. Male fiddler crabs fighting over possession of a burrow never injure each other although the enlarged claws which they use are powerful enough to crush the abdomens of their opponents. Rattlesnakes often wrestle with each other but never bite; male deer lower their heads and lock their branched antlers together but seldom use them to pierce each other's bodies; and some antelopes actually kneel down when engaging in combat. Many other examples could be given (see e.g. Lorenz, 1966) of animal conflicts which are settled by ritualized displays without any physical contact at all.

Although escalated fighting is by no means unknown in the animal kingdom (Geist, 1966), or for that matter among humans, conventional fighting is sufficiently common to require an explanation. The problem is this: animals do not use their weapons to maximal effect in conventional combats. In a contest between two animals, A and B, if A adopts a conventional fighting strategy and B an escalated strategy, it would seem that B would gain the advantage and pass more of its genes on to its offspring, on average, than A. How, then, can conventional fighting be explained in the framework of natural selection?

Non-strategic explanations (e.g. Lorenz, 1966; Eibl-Eibesfeldt, 1970) are traditionally based on appeals to "the good of the species". A typical argument runs as follows. If the members of a population were to engage in escalated fighting, serious injuries would be common and the survival of the population as a whole would be placed in jeopardy. A population in which conventional fighting is the norm has a better chance of survival. Natural selection has therefore favoured conventional fighting.

This argument is based on the hidden assumption of *group selection*: it is assumed that the fittest *populations* (rather than the fittest individuals) survive in the process of natural selection. Group selectionist arguments, however, are regarded with suspicion by most contemporary evolutionary theorists (see the review by Maynard Smith, 1976b). The fundamental theoretical problem is to explain how a population of conventional fighters could arise in the first place. Imagine a population of escalated fighters in which a mutation occurs causing a few animals to fight conventionally. Provided that the advantage of winning outweighs the risk of injury attendant upon ecalated combat, the mutants would fare badly against the majority in terms of Darwinian fitness and their genes would soon be eradicated by selective pressures operating *within* the population. In a population of conventional fighters, on the other hand, a mutation which caused escalated fighting would spread through the population because of the increased Darwinian fitness of its possessors even if it reduced the chances of survival of the population as a whole.

The group selectionist argument is therefore unconvincing except in certain very unusual circumstances outlined by Maynard Smith (1976b). What is required is an explanation of conventional fighting showing how it increases the Darwinian fitness of the *individuals* displaying it. This requirement is fulfilled by Maynard Smith and Price's (1973) application of game theory to the problem.

Suppose that there are two pure fighting strategies available to each individual: Hawk and Dove. The Hawk (H) strategy involves escalated fighting until injury forces withdrawal or the opponent gives way. The Dove (D) strategy involves conventional fighting; the individual adopting it retreats before getting injured if its opponent escalates. An animal's genotype, we shall assume, determines the strategy that it adopts.

After each contest the animals receive their payoffs. The expected payoff to a Hawk in a contest with a Dove is written $E(H, D)$, the expected payoff to a Dove in a contest with another Dove is $E(D, D)$, and so on. The payoff is a measure of the change in Darwinian fitness of the contestant, that is, the increase or decrease in its expected number of offspring, and it depends on three factors. The first is the advantage of winning: the resource over which the contest takes place is assumed to be worth V (for "victory") units of Darwinian fitness to the winner. The second is the disadvantage of being injured: injury alters an animal's fitness by $-W$ (for "wounds") units. The third is the time and energy wasted in a long contest, which alters the fitness of each contestant by $-T$ (for "time") units in conventional fights.

The payoffs associated with every possible combination of strategies can be worked out quite easily from these components. If a pair of contestants both adopt Hawk strategies, we may assume that, on average, their chances of winning are equal, their chances of being injured are also equal, and the contest will be short so neither will waste unnecessary time and energy; an individual's expected payoff will therefore be $1/2(V - W)$. If both adopt Dove strategies, their chances of winning are once again equal, but there will be no chance of injury to either, and the contest will be a long one in which time and energy are wasted by both contestants; an individual's expected payoff will therefore be $1/2V - T$. A Dove in conflict with a Hawk will flee as soon as the latter escalates and will therefore receive a zero payoff. Finally, a Hawk in conflict with a Dove will win the resource without risk of injury or wastage of time and energy, yielding a payoff of V.

Matrix 12.1a summarizes these payoffs in the form of a generalized payoff matrix for the Hawk–Dove game as given by Maynard Smith and Price (1973) and Maynard Smith (1976a). Matrix 12.1b shows a possible set of numerical payoffs based on the following assumed values suggested for illustrative purposes by Maynard Smith (1978a): $V = 10$, $-W = -20$, and $-T = -3$.

Matrices 12.1
The Hawk–Dove Game

		II				II	
		H	D			H	D
I	H	$1/2\ (V - W)$	V	I	H	-5	10
	D	0	$1/2\ V - T$		D	0	2
		(a)				(b)	

Note that the payoffs shown in the matrices are those to Player I; the payoffs to Player II are normally omitted from evolutionary game matrices although the strategic structures are not zero-sum. The games are, however, the same from both players' points of view, so Player II's payoffs can easily be filled in if required.

Maynard Smith and Price (1973) introduced the concept of an *evolutionarily stable strategy* to handle games of this sort. Suppose that the members of a population play the Hawk–Dove game in random pairs and that the animals then produce offspring which use the same strategies as their parents. The following question can then be asked. Is there a strategy with the property that if most members of the population adopt it, no mutant strategy can invade the population by natural selection? Such a strategy is called an evolutionarily stable strategy, or ESS for short, because no mutant strategy confers a higher Darwinian fitness on the individuals adopting it and the ESS is consequently invulnerable to invasion by the available alternative strategies. It is therefore the strategy we would expect to see commonly in nature.

The game shown in Matrix 12.1b turns out, on examination, to be a game of Chicken (see Section 6.5). The cautious minimax strategy is Dove, and unilateral defection from it benefits the defector and harms the minimax player, while joint defection results in the worst possible payoffs for both players. The game does not possess a symmetric equilibrium point in pure strategies, and a consequence of this is that neither of the pure strategies is an ESS. This can be shown quite easily. A population of Hawks can be invaded by Doves because the payoff to a Hawk in such a population is $E(H, H) = -5$ while the payoff to a Dove is $E(D, H) = 0$. It follows that a

Dove mutant would produce more offspring than a Hawk. A similar argument can be used to show that the Dove strategy is not an ESS: in a population of Doves a Hawk mutant would produce more offspring than a Dove.

Maynard Smith (1974, 1978a, 1978b) has specified the mathematical requirements of an ESS. (The computations are extremely cumbersome, but a much simpler analysis is given in the following section of this chapter.) Suppose a population consists mostly of individuals adopting Strategy I but a small fraction p of mutants adopt Strategy J. An individual adopting Strategy I will receive a payoff of $E(I, I)$ with probability $1 - p$ and a payoff of $E(I, J)$ with probability p. The payoff to an individual adopting Strategy J will be $E(J, I)$ with probability $1 - p$ and $E(J, J)$ with probability p. If the Darwinian fitness of each member of the population before a series of contests is C, then after the contests the fitness of an individual adopting Strategy I, denoted by $W(I)$, will be

$$W(I) = C + (1 - p)E(I, I) + pE(I, J),$$

and the fitness of an individual adopting Strategy J, denoted by $W(J)$, will be

$$W(J) = C + (1 - p)E(J, I) + pE(J, J).$$

If I is an ESS, then by definition $W(I) > W(J)$. Since p is assumed to be small, this implies that

either $E(I, I) > E(J, I),$ (1)

or $E(I, I) = E(J, I)$ and $E(I, J) > E(J, J).$ (2)

Conditions (1) and (2) are Maynard Smith's definitions of an ESS.

In the Hawk–Dove game shown in Matrix 12.1b, the ESS turns out to be a mixed strategy of $8/13H$ and $5/13D$. Either 8/13 of the population will consistently adopt the Hawk and 5/13 the Dove strategy, or every individual will play Hawk 8/13 and Dove 5/13 of the time. To prove that this is an ESS it is necessary to show that it is invulnerable to invasion by either Hawks or Doves.

Denoting the mixed strategy by M it can be shown first that

$$E(M, M) = E(H, M)$$

and $E(M, H) > E(H, H),$

confirming that M is evolutionarily stable against H according to Definition (2) above. The calculation is as follows:

$$E(H, M) = 8/13E(H, H) + 5/13E(H, D),$$
$$E(D, M) = 8/13E(D, H) + 5/13E(D, D),$$
$$E(M, H) = 8/13E(H, H) + 5/13E(D, H),$$
$$E(M, M) = 8/13E(H, M) + 5/13E(D, M).$$

The values of $E(H, H)$, $E(H, D)$, $E(D, H)$, and $E(D, D)$ are the numbers in Matrix 12.1b; they are -5, 10, 0, and 2 respectively. Substituting these values in the above equations, we arrive finally at

$$E(M, M) = 10/3 = E(H, M)$$

and $$E(M, H) = -40/13 > E(H, H) = -5,$$

which establishes that M is evolutionarily stable against H according to Definition (2). The same method of analysis can be used to show that M is evolutionarily stable against D. In other words, since H and D are by hypothesis the only pure strategies available, M is an ESS in the Hawk–Dove game shown in Matrix 12b.

In the generalized Hawk–Dove game shown in Matrix 12.1a, in which no assumptions are made about the fitness equivalents of victory (V), wounds ($-W$), and time and energy wastage ($-T$), the algebraic conditions which an ESS must satisfy are as follows.

For Hawk to be a pure ESS according to Definition (1), it must be the case that $E(H, H) > E(D, H)$. Substituting the expressions given in Matrix 12,1a, we have

$$1/2(V - W) > 0,$$

i.e. $$V > W.$$

Alternatively, according to Definition (2), Hawk is a pure ESS if $E(H, H) = E(D, H)$ and $E(H, D) > E(D, D)$. Substituting as before, we have

$$1/2(V - W) = 0,$$

i.e. $$V = W,$$

and $$V > 1/2V - T.$$

The inequality immediately above is necessarily true since V and T are both positive numbers.

By a similar method of calculation—working straight from the definitions—it can be shown that Dove can never be a pure ESS: the definitions cannot be satisfied for any values of V, W, and T. The only other possibility is a mixed ESS, and by elimination this must be the result when $V < W$. To summarize, Hawk is a pure ESS when $V \geq W$, and there is a mixed ESS when $V < W$.

All of this looks (and is) unnecessarily complicated. The awkwardness arises from Maynard Smith's use of a two-person game to model what is, in fact, a multi-person strategic interaction. A more natural and straightforward analysis is presented in the following section.

12.4   An Improved Multi-Person Game Model

A natural way of modelling the Hawk–Dove problem is to construct a compound multi-person game in which the players engage in a series of two-person contests with one another according to the payoff structure of Matrices 12.1. The general theory of compound games, outlined in Section 8.8, can be adapted to the problem as follows.

   If each animal adopts either a Hawk or a Dove strategy (or a mixture of the two) in a series of contests with other animals, its expected payoff will be a linear function of the proportion of other members of the population playing Hawk or Dove. It does not make any difference whether each animal plays with all the others or whether each encounters a random sample of opponents, but it is assumed that all animals are, on average, involved in the same number of contests. Since we are concerned only with the *relative* fitness of the genotypes, we can ignore the total number of contests and concentrate instead on the expected payoffs per contest.

   Suppose the members of a population play the two-person game shown in Matrix 12a in pairs. The expected payoff to an individual in each contest depends on its own strategy and the proportion of other animals in the population that adopts the Hawk strategy. Let this proportion be k; the proportion of Dove opponents is then $1 - k$. No restrictions are placed on the proportion of Hawks: k can vary from zero to unity. If an individual plays Hawk, then its expected payoff, written $E(H)$, is simply

$$E(H) = 1/2(V - W)k + V(1 - k),$$

and if it plays Dove, its expected payoff $E(D)$ is

$$E(D) = (1/2V - T)(1 - k).$$

With the values shown in Matrix 12.1b, the resultant compound game is depicted in Figure 12.1.

   When the Hawk–Dove problem is modelled by a compound game in this way, an ESS is simply an equilibrium point in the game. It is possible to see at a glance that there is no pure ESS. In a population consisting mostly of Doves (when k is small at the left of the graph), the expected payoff to a Hawk $E(H)$ is higher than the expected payoff to a Dove $E(D)$. In such a population Hawks will therefore produce more offspring than Doves and the proportion k of Hawks will increase by natural selection, shifting the outcomes to the right until the point is reached at which the $E(H)$ and $E(D)$ payoff functions intersect. At this point the payoffs to Hawks and Doves are equal and the proportions will remain stable. In a population consisting mostly of Hawks (when k is large towards the right of the graph), the expected payoff to a Dove $E(D)$ is higher than the expected payoff to a Hawk $E(H)$, so the proportion k of Hawks will decrease until the intersec-

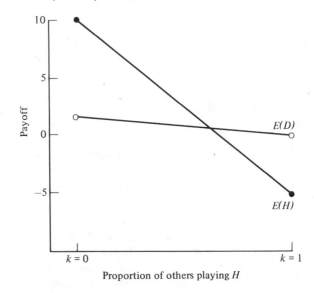

FIG. 12.1   A compound version of the Hawk–Dove game shown in Matrix 12.1b. The vertical coordinate represents an animal's expected payoff in units of Darwinian fitness, and the horizontal axis represents the number of other members of the population playing Hawk. The $E(H)$ function shows the expected payoff to the animal if it plays Hawk, and the $E(D)$ function shows its expected payoff if it plays Dove.

tion point is reached. It is clear that, whatever the starting proportions of Hawks and Doves, the population will evolve to the ESS equilibrium point at the intersection of the payoff functions.

What is the proportion of Hawks at the ESS equilibrium point? Since the payoffs to a Hawk and a Dove are equal where the payoff functions intersect, the value of k can be found very simply by equating the functions and solving for k. With the figures shown in Matrix 12.1b, for example, the payoff functions $E(H)$ and $E(D)$ are equated as follows:

$$-5k + 10(1 - k) = 2(1 - k),$$
$$k = 8/13,$$

indicating an equilibrium when 8/13 of the population (strictly speaking, the population minus one individual) are Hawks. This confirms the laborious calculation given (in part) in the previous section. It is possible, of course, to interpret the mixed ESS to mean that each individual plays Hawk 8/13 of the time.

This compound game model has several advantages over the two-person models used by Maynard Smith and his colleagues. In the first place, the computation of ESSs is greatly faciiitated, as the Hawk–Dove example shows. Secondly, the model clarifies the underlying strategic structure of the evolutionary process by showing that it is a multi-person game involving a whole population of players and that the critical parameter is k, the relative frequency of a particular genotype in the population. Thirdly, the directions of evolutionary change from different starting conditions (values of k) can be seen at a glance. Lastly, it is possible to provide a typology of binary (two-strategy) evolutionary games, revealing the full range of theoretical possibilities and generating testable hypotheses, as will be shown presently. Multiple-strategy evolutionary games can be dealt with by breaking them down into their binary constituent parts.

Matrix 12.2

	A	B
A	w	x
B	y	z

A generalized payoff matrix for a binary evolutionary game is shown in Matrix 12.2. The expected payoff to an A type in competition with another A type is w units of Darwinian fitness, the expected payoff to an A type in competition with a B type is x units, and so on. If the members of the population compete with one another in random pairs, either simultaneously or sequentially, the resultant compound multi-person game can be represented graphically by means of a pair of payoff functions. The expected payoff to an A type is written $E(A)$ and the expected payoff to a B type is $E(B)$. When the proportion of A types in the population is k, the expected payoff to an A type is

$$E(A) = wk + x(1 - k),$$

and the expected payoff to a B type is

$$E(B) = yk + z(1 - k).$$

The equation of the $E(A)$ payoff function defines a straight line which extends from a payoff of x when $k = 0$ to a payoff of w when $k = 1$. The $E(B)$ function is a straight line from z when $k = 0$ to y when $k = 1$.

There are eight non-degenerate types of cases, shown diagrammatically in Figure 12.2.

(i) The $E(A)$ function lies above the $E(B)$ function throughout its length. In this case Strategy A dominates Strategy B, so the ESS is A.

(ii) The $E(B)$ function lies above the $E(A)$ function throughout its length. The ESS is B.

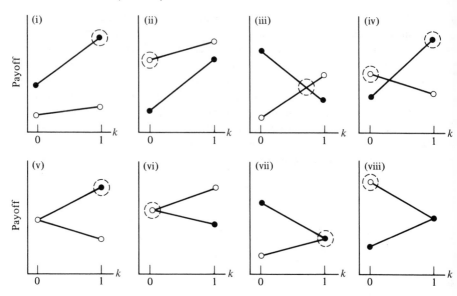

FIG. 12.2 The eight qualitatively different types of compound multi-person evolutionary games. The vertical axes represent an individual's expected payoffs in units of Darwinian fitness, and the horizontal axes represent the proportions k of A types in the population. The expected payoffs to an individual A type are shown by the lines connecting filled dots and the expected payoffs to an individual B type by the lines connecting open dots. Evolutionarily stable compositions are indicated by dashed circles.

(iii) The functions intersect with $E(A) > E(B)$ when $k = 0$ and $E(A) < E(B)$ when $k = 1$. In this case the ESS is a mixed strategy located at the intersection point. It can be found by solving the following equation (derived by equating the payoff functions and simplifying):

$$k = (z - x)/(w - x - y + z).$$

(iv) The functions intersect with $E(A) < E(B)$ when $k = 0$ and $E(A) > E(B)$ when $k = 1$. There are two pure ESSs in this interesting case, and the final result depends on the starting condition. The population will evolve to the A ESS when $k > (z - x)/(w - x - y + z)$ and to the B ESS when $k < (z - x)/(w - x - y + z)$.

(v) $E(A) = E(B)$ when $k = 0$, and $E(A) > E(B)$ when $k = 1$. The ESS is A.

(vi) $E(A) = E(B)$ when $k = 0$, and $E(A) < E(B)$ when $k = 1$. The ESS is B.

(vii) $E(A) = E(B)$ when $k = 1$, and $E(A) > E(B)$ when $k = 0$. The ESS is A.

(viii) $E(A) = E(B)$ when $k = 1$, and $E(A) < E(B)$ when $k = 0$. The ESS is B.

There is at least one ESS in every case. A mixed ESS occurs only in one case (iii), and in one other (iv) there are two pure ESSs. The model can be used to compare a specified genotype with several mutant forms, one at a time, in order to see whether populations composed in various ways will evolve to new ESSs. No genotype is evolutionarily stable in an absolute sense since it is impossible to conceive of all the mutant forms that might arise. But an existing population, pure or mixed, can be examined for stability against invasion by a single specified mutant form, and this analysis can be repeated for other specified mutants. In most cases the order in which the comparisons are made is immaterial. It occasionally happens, however, that A dominates B, B dominates C, and C dominates A. If A, B, and C are all present in the population, endless cycling will occur without an ESS ever being attained. This is closely related to Condorcet's paradox, discussed at length in Sections 10.5 and 10.6. An example of such a game without an ESS is given in the appendix to Maynard Smith (1974).

Maynard Smith and Price (1973) analysed a number of animal conflict games in which more than two pure strategies are available to each player. One of their best known models is the Hawk–Dove–Retaliator game. This is an extension of the Hawk–Dove game, with the additional option available to an individual animal of fighting conventionally and escalating only if its adversary escalates. In other words a Retaliator normally plays Dove but responds to a Hawk opponent by playing Hawk. Maynard Smith and Price's analysis is based on a two-person game with three rows and three columns, and it is possible to show that the ESS is Retaliator. This can be shown more easily with the help of the compound game model.

Suppose that the proportion of Hawks in a population is h, the proportion of Doves is d, and the proportion of Retaliators is r, with $h + d + r = 1$. The expected payoff to a Hawk in a contest with another Hawk is evidently $E(H, H)$, and if opponents are encountered at random the probability that a Hawk will receive this payoff is h since the proportion of potential Hawk opponents in the population is h. The expected payoff to a Hawk in a contest with a Dove is $E(H, D)$ and its corresponding probability of occurrence is d. Retaliators adopt the Hawk strategy against Hawks, so the expected payoff to a Hawk in a contest with a Retaliator is $E(H, H)$ occurring with probability r. In a contest with an opponent chosen at random from the population, the overall expected payoff $E(H)$ to a Hawk is therefore $E(H, H)h + E(H, D)d + E(H, H)r$. It can be shown analogously that a Dove's expected payoff $E(D)$ in a random encounter is $E(D, H)h + E(D, D)d + E(D, D)r$, and a Retaliator's expected payoff $E(R)$ is $E(H, H)h + E(D, D)d + E(D, D)r$.

Substituting the values of $E(H, H)$, $E(H, D)$, $E(D, H)$, and $E(D, D)$ shown in Matrix 12.1a, the Hawk, Dove, and Retaliator payoff functions can be rewritten as follows:

$$E(H) = 1/2(V - W)h \qquad\qquad + Vd + 1/2(V - W)r,$$
$$E(D) = \qquad\qquad 0h + (1/2V - T)d + (1/2V - T)r,$$
$$E(R) = 1/2(V - W)h + (1/2V - T)d + (1/2V - T)r.$$

Each of the three fighting strategies can now be tested for evolutionary stability as follows. The Hawk strategy is an ESS if, when most of the population adopt it, rare Dove or Retaliator mutants cannot invade the population and spread through it at the expense of the Hawks. If most of the population adopt the Hawk strategy, then h is close to 1 and d and r are close to zero. We can accordingly compare the payoffs to Hawks, Doves, and Retaliators as h approaches 1 while d and r approach zero. Under these conditions it is obvious by inspection of the payoff functions that $E(D) > E(H)$ provided that $V < W$ as in the numerical example of Matrix 12.1b. With this constraint, therefore, if most of the population are Hawks, Dove mutants will multiply faster than the Hawks, so the Hawk strategy is not an ESS.

To see whether the Dove strategy is an ESS, we compare the payoffs to Hawks, Doves, and Retaliators as d approaches 1 while h and r approach zero. In this case it is clear that $E(H) > E(D)$; this means that Hawk mutants will multiply faster than Doves, so the Dove strategy is not an ESS.

We turn finally to the Retaliator strategy. As r approaches 1 while h and d approach zero, neither $E(H)$ nor $E(D)$ is greater than $E(R)$ provided that $1/2W > T$ as in the numerical example of Matrix 12.1b. Under this constraint, therefore neither Hawk nor Dove mutants will multiply faster than Retaliators, so the Retaliator strategy is an ESS.

We may conclude that, with the specified constraints on the relative values of V, W, and T, the ESS of the Hawk–Dove–Retaliator game is the Retaliator strategy. A similar conclusion was reached by Maynard Smith and Price (1973) via much more difficult calculations. A complete solution of the game is as follows. If $V < W$ and $1/2W > T$, then Retaliator is the only ESS; but if $V \geq W$, then Hawk is also an ESS, and if $1/2W \leq T$, then Retaliator is not an ESS.

The compound game model suggests some refinements of the definition of an ESS. Maynard Smith (1978a) defines an ESS as "a strategy with the property that if most members of a large population adopt it, then no mutant strategy can invade the population" (p. 142). But this is not entirely clear. To begin with, an ESS appears to be defined in an all-or-none fashion, whereas the underlying concept of evolutionary invasion is evidently a matter of degree. In the Hawk–Dove game analysed earlier, in which the ESS is mixed, if the population consists mostly of Hawks, can Dove mutants invade it? The answer seems to be that Doves can *partly* invade: the invasion proceeds until 5/13 of the population are Doves, and the extent of invasion could be made arbitrarily small by a suitable choice of payoff functions.

Secondly, if a mixed ESS consists of some members of the population always playing Hawk and the rest always playing Dove, then, strictly speaking, there is no ESS according to Maynard Smith's definition: there is no strategy with the property that if *most* adopt it no mutant can invade the population. If most adopt Hawk in a Type (iii) game, Doves invade (up to a point), if most adopt Dove, Hawks invade (up to a point), and no other strategy is ever played.

Thirdly, there is one type of binary compound game (iv) in which a population containing A type individuals will be completely invaded if and only if the proportion of A type individuals falls below a critical value. Thus A is stable against B unless the proportion playing A falls below the value of k at the intersection point; when the proportion is smaller than this the population evolves to B. In Figure 12.2(iv), if "most" of the population are B types—with k less than 1/2 but not as small as at the intersection point—then A types will invade the population and lead to a pure A ESS. The intersection point can, of course, be arbitrarily close to either of the end-points of the graph. The classical definition of an ESS is inappropriate in these cases, first because there are two pure ESSs, and secondly because one of them may be an ESS in spite of the fact that "if most members of a large population adopt it" another strategy can invade the population and replace it.

It would seem preferable to define evolutionary stability as a property of the *composition* of a population, rather than as a property of a *strategy*. An evolutionarily stable composition can be defined informally as one that tends to regain its prior proportions if perturbed. Strictly speaking, however, one should specify the strategies to which this stable composition refers. An evolutionarily stable composition (ESC) of Strategies A and B is one in which either

(a) if the proportion of A types in the population increased marginally, then A types would not be fitter than B types; or
(b) if the proportion of A types decreased marginally, then B types would not be fitter than A types.

This definition of an ESC can be formalized as follows. Suppose a proportion k of the population adopt Strategy A and the remainder adopt Strategy B. The expected payoff to each individual adopting A in this contingency is written $E(A, k)$ and the expected payoff to each individual adopting B is $E(B, k)$. Now suppose that k increases by an arbitrary small amount t or decreases by an arbitrary small amount $-t$. Then k is an ESC if

either $$E(A, k + t) \leqslant E(B, k + t),$$

or $$E(B, k - t) \leqslant E(A, k - t).$$

This concept of an ESC seems to avoid many of the problems associated with the original ESS concept.

12.5   Empirical Evidence

A number of explicit and implicit simplifying assumptions are built into the game models described above and into others like them (Maynard Smith, 1978b; Parker, 1978). Whether or not the models provide adequate accounts of strategic evolutionary processes in nature is therefore ultimately an empirical question. The evidence is, however, rather scanty for two main reasons. First, it is extremely difficult to design a stringent test of an evolutionary game model because the payoffs, although fully numerical and rigorously defined, are not normally known in specific instances. Investigators have usually had to rely on indirect and partial indices of Darwinian fitness such as the number of eggs laid or the number of matings observed at a certain time and place.

A second reason for the relative sparseness of the empirical evidence is that no-one sought the relevant data until the pertinent questions were posed by theoretical analyses in the 1970s. Maynard Smith (1978b) has remarked that "there is a real danger that the search for functional explanations in biology will degenerate into a test of ingenuity. An important task, therefore, is the development of an adequate methodology of testing" (p. 52). Such evidence as there is has been reviewed by Parker (1978), Lazarus (1982) and others; in this section only a few of the more striking examples are discussed.

A fairly weak prediction of the Hawk–Dove–Retaliator model is that retaliation ought to be a common feature of animal combat since a population of Retaliators is evolutionarily stable against Hawk and Dove mutants. The available evidence (reviewed by Geist, 1974, 1978) confirms this prediction which is, however, consistent with various other possible models. The following example is typical: a rhesus monkey that loses a contest over a resource will passively submit to relatively harmless incisor bites from the victor, but it will retaliate with great ferocity if the victor uses its canines (Bernstein and Gordon, 1974).

Several examples of populations with mixed strategies (or mixtures of pure strategies) have come to light, and in a few cases there is evidence of evolutionary stability. Parker (1970) provided such evidence in a study of male dung flies of the genus *Scatophaga*. The females lay their eggs in cow pats, so groups of males tend to hover around cow pats in order to meet females with which to mate. A stale cow pat attracts fewer females than a fresh one. When a cow pat becomes stale, a male is therefore faced with a straightforward strategic choice: he can leave early in search of females elsewhere or he can stay, and his payoff, indexed by the expected number of female captures, clearly depends on the number of other males that leave early. If most of the other males leave early, he is better off staying since there will be very little competition for the females that arrive even though the females will be relatively few in number. But if most of the other males stay, he does better by leaving early in search of females elsewhere. The corresponding game is Type (iii) in Figure 12.2, and the ESS or ESC is therefore a mixture in which some males leave early and some stay.

According to game theory, early leavers and stayers should receive the same payoffs, that is, the same average number of female captures. Parker (1970) showed that the predicted and observed distributions of the stay times of male flies, based on 15,942 field observations, were in extremely close agreement. He also showed—and this is a much more stringent test of the theory—that the average frequency of female captures was almost identical for males that left early and for those that remained on the cow pat (see also Parker, 1978).

Very clear evidence for the existence of a mixed ESS or ESC was provided by Brockmann, Grafen, and Dawkins (1979) in a study of the great golden digger wasp, *Sphex ichneumoneus*. The males of this species, apart from mating with the females, spend most of their time drinking nectar from flowers. The females dig underground burrows in the summer and provision them with grasshoppers which they have captured and paralysed. The grasshoppers provide food for the wasps' offspring when they hatch. Each female lays a single egg in each burrow and seals it if she gets a chance to do so, but ant invasions and other natural disasters often force her to abandon a burrow, for a while at least, before laying an egg in it.

An individual female therefore has two alternative strategies: to dig her own burrows or to enter burrows abandoned by other females. It is clearly better for her to enter burrows prepared by others if most of the others dig their own burrows, but equally clearly some of the females (or all of them on some occasions) have to do the work of preparing burrows. In other words the underlying strategic structure of the problem is once again of Type (iii) and a mixture of digging and entering is predicted by the theory. Among 30 female wasps in Brockmann's New Hampshire study area a mixture of digging and entering was indeed observed. This was shown to be a probable ESS or ESC since the rates of egg laying, which provide a reasonable index of payoffs, were very similar for the two strategies.

An interesting example of the Concorde fallacy turned up in the same species (Dawkins and Brockmann, 1980). The Concorde fallacy, discussed in Section 8.6, is the tendency to continue investing in a project on account of past expenditure on it rather than expected future payoff, and female digger wasps often commit it.

A female cannot distinguish between an abandoned burrow and one whose owner is away on a provisioning trip. Occasionally two females meet at the same burrow, and the consequence is invariably a fight which continues until one of the wasps is driven off. The prize to the victor is a burrow, which may already be well provisioned with grasshoppers, in which to lay an egg. How hard ought a wasp to fight over a disputed burrow? If the burrow is empty it is hardly worth fighting for; she would be better off giving up and spending a few hours digging a new one. But if the burrow is already well provisioned it is worth fighting harder since many days of hunting will be saved by gaining possession of it. The larger the total prize, the greater the justification for fighting over it.

Dawkins and Brockmann (1980) observed 23 fights, and were surprised to discover that the length of time a wasp was prepared to fight depended not on how much food there was in the burrow, but on how much she herself had contributed to it. The wasp that had carried the least amount of prey into the burrow was generally the first to give up fighting. This is a striking example of the Concorde fallacy, as the authors pointed out.

It would seem that the wasps' behaviour is not evolutionarily stable against a mutant form that fights according to the total contents of the burrow. A wasp that regulated its fighting in this improved manner would lay more eggs, on average, than one that commits the Concorde fallacy. Why, then, has such a mutant form not invaded the population? The answer may lie in the limited intellectual capacities of the great golden digger wasp, which possibly ensure that the improved behavioural pattern is not an available alternative. A wasp may simply be unable to count the number of grasshoppers in a burrow, although she can evidently judge how many grasshoppers she herself has carried there. Natural selection can operate only on an existing gene pool, and these wasps may not have the necessary genes for evolving an optimal strategy. In this case the observed composition of the wasp population is presumably an ESC, defined in relation to the available alternative strategies, until the necessary mutations occur. Another species of digger wasp (*Ammophila campestris*) is known to have the requisite genes.

A number of experimental tests of ESC theory suggest themselves. In populations of plants or animals in which two (or more) clearly distinguishable alternative genotypes, A and B, are present, the relative payoffs at the endpoints of the graphs could be estimated by forming isolated breeding populations with artificially extreme compositions (mostly A and mostly B) and comparing the reproductive successes of the two types during the first few generations. Once these estimates have been made, the ESC can be predicted; it will be a uniform composition of A types or a uniform composition of B types, as shown in Figure 12.2, except in cases falling into Type (iii). In Type (iii) cases the composition of the experimental population should eventually stabilize at a value of k similar to that found in the wild population, since the ESC is independent of the number of individuals (ignoring rounding errors in very small groups in which certain exact proportions are unattainable—for example three-quarters of a three-person group cannot be female).

Species in which different isolated breeding populations in the wild are known to exist in different uniform compositions (all A in one population and all B in another) are strongly suggestive of Type (iv) games. By mixing experimental populations in predetermined ways, that is, by choosing different values of k, it may be possible to demonstrate two opposite processes of evolutionary change to different ESCs.

These experiments may, however, present difficulties of ensuring that environmental conditions in the experimental populations that affect the payoffs are similar to those in the wild.

12.6 Summary

The chapter began with a brief historical outline, from 1930 to the present, of applications of game theory to evolution. The most important contributions began with Maynard Smith's work in the 1970s. In Section 12.2 the existence of strategic processes in evolution was pointed out and it was shown how the concepts of game theory can be placed in one-to-one correspondence with those of natural selection: players correspond to individual organisms, strategies to genotypes, and payoffs to changes in Darwinian fitness. The essential ideas were illustrated with reference to the evolution of sex ratios.

In Section 12.3 the problem of conventional fighting in animals was raised. Why do animals sometimes avoid using their weapons to maximal advantage? Group selection provides a dubious explanation, but game theory explains how conventional fighting can arise in some members of a population, or in all members some of the time, by individual selection alone: Maynard Smith's two-person Hawk–Dove game can have a mixed evolutionarily stable strategy. An improved compound multi-person game model was outlined in Section 12.4. This led to a typology of evolutionary games and to a new concept of an evolutionarily stable composition of a population. In Section 12.5 some relevant empirical evidence was outlined. Retaliation, which ought to be a common feature of animal combat according to the theory, is often observed in practice. Evidence for evolutionarily stable behavioural mixtures has been found in dung flies and in digger wasps. One species of digger wasp is prone to the Concorde fallacy, but this does not necessarily refute the theory. Some new experimental methods of testing the theory were suggested.

13

Moral Philosophy and Practical Problems of Strategy

13.1 Game Theory and the Conduct of Life

THE PUBLICATION of von Neumann and Morgenstern's *Theory of Games and Economic Behavior* in 1944 was greeted with unprecedented enthusiasm by those who thought they saw in it a method of solving virtually all strategic problems of everyday life (see, for example, the review of von Neumann and Morgenstern's book by Copeland, 1945). But the hard truth of the matter is that formal game theory seldom provides straightforward solutions to practical problems. There are three major reasons for this.

(a) Persuasive formal solutions exist only for strictly competitive (two-person, zero-sum) games, but real strategic interactions are not usually strictly competitive.

(b) Human beings have bounded rationality and cannot be expected to analyse the payoff matrices of any but the simplest games. Chess, for example, is a strictly competitive game, no more complex than many everyday strategic problems, and the minimax theorem tells us that it has a formal solution; but knowing this is of no help in actually playing the game or even in programming a computer to play it. The winner of the Third World Computer Chess Championship in 1981, known as Belle, conducts a minimax search for the "best" move according to certain restricted criteria by evaluating 160,000 possible positions per second, or about 29 million for each move it makes; but this "brute force" search enables it to "think" only a few moves ahead, and Belle plays considerably below grandmaster strength (*Scientific American*, 1981). The information-processing capacity of human beings is, of course, much more severely limited, and "brute force" calculation is out of the question in most positions, so human players rely largely on positional judgment.

(c) Two characteristic features of everyday strategic interactions are, first, that the rules and payoff functions change—and are often deliberately

254

changed by the players—while the interactions are in progress (Colman, 1975), and secondly that the players have incomplete information about the games they are playing. These features can be incorporated into game theory at a formal level (see, e.g., Brams, 1977; Mertens and Zamir, 1980) but only at the price of vastly increased complexity, making the theory as a guide to practical action all the more vulnerable to (b).

How, then, do human beings handle complex strategic problems, or how ought they to handle them? The most influential theory bearing on this question was originally formulated by Herbert Simon (1957, pp. 241–73, 1976) to explain organizational decision making. Rejecting the approach of neoclassical economics, according to which decision makers act to optimize their payoffs, Simon put forward the concept of *satisficing*. Instead of searching for the optimum strategy in a situation, a process that consumes an undue amount of time and energy and is often, in any event, impossible to complete, the wise decision maker, according to this theory, searches just long enough to find a strategy that is satisfactory, or that suffices; in other words, one that satisfices. A couple looking for a house to buy, for example, usually settle for the first one to come along that is acceptable according to certain minimal requirements—price, location, number of rooms, amenities, and so on—without attempting to examine every single available alternative to ensure that their choice is optimal. Human chess players also usually adopt satisficing moves, something that computers cannot be programmed to do effectively on account of their lack of positional judgment. Belle, in common with other leading chess programs, is weakest in what chess players call "simple" positions, where positional judgment is at a premium.

Strategic choices in everyday life, since they affect other people, often raise moral problems, and game theory can sometimes help to elucidate these problems without "solving" them in a formal sense. The first moral philosopher to apply game theory in this area was Richard Braithwaite (1955), who had the following to say:

> The algebra I enlist for this purpose is the new mathematical Theory of Games; and I hope, by showing how it can yield results, to encourage others as well as myself to pursue and apply it further. . . . Perhaps in another three hundred years' time economic and political and other branches of moral philosophy will bask in radiation from a source—the theory of games of strategy—whose prototype was kindled round the poker tables of Princeton (pp. 54–5).

Joseph Fletcher (1966), the originator of "situation ethics" was equally sanguine: "Moral choices need intelligence as much as they need concern. . . . To be 'good' we have to get rid of innocence . . . [perhaps] by learning how to assign numerical values to the factors at stake in problems of conscience" (pp. 114, 118).

It has been argued, most forcefully by Martin (1978), that game theory has various built-in assumptions which predispose it to handle certain kinds of problems and to generate certain kinds of conclusions. There is undoubtedly a measure of truth in this argument, although it is worth bearing in mind that it applies equally to any other abstract framework such as probability theory or statistics. According to Martin, it follows that "applying a game theory framework to ethical situations is more likely to obscure satisfactory solutions [such as changing the rules] than reveal them" (p. 98). But this would appear to be prejudging the issue: the usefulness or otherwise of a theory when applied to a particular class of problems is best determined by trying it out in practice.

If we accept the usual distinction of social actions into means and ends, perhaps the major limitation of game theory when applied to ethical problems is that it takes the players' ends as given and examines only the rationality of particular means to achieve these ends; in the terminology of Max Weber (1921) it deals only with *Zweckrationalität* to the exclusion of *Wertrationalität*. Rationality as conceived by eighteenth-century rationalist philosophers, and by many present-day non-philosophers, is concerned partly with the ends themselves (*Wertrationalität*). Game theory is completely neutral with regard to the players' systems of preferences (payoff functions) provided that they are internally consistent, but it can draw certain conclusions about rational choice once these preferences are specified. The predilections and limitations of game theory have been discussed in more detail by Schelling (1968) in his essay on "game theory and the study of ethical systems".

Applications of game theory to ethical problems fall into three broad classes. The first class of applications has its origins within game theory itself: a number of models have been developed to define arbitration schemes or bargaining "solutions" to simple abstract games which are not strictly competitive. These models are generally aimed at establishing outcomes that are "fair" and workable at a purely formal level. Braithwaite's (1955) contribution is of this kind, as is the work of Shapley (1953) referred to in Chapter 8; but these applications will not be pursued here since, for reasons outlined earlier, they have limited relevance to practical problems.

A second class of applications involves the use of game theory to elucidate existing issues in moral philosophy. Among the issues to which game theory has had something to contribute are the nature and ethical implications of rationality and the pursuit of self-interest (discussed below in Section 13.2), Kant's categorical imperative (Section 13.3), and Rousseau's social contract theory (Section 13.4).

In the third and perhaps the most interesting class of applications are entirely new problems of moral philosophy which were discovered by investigating the implications of game theory in particular cases. These problems were not recognized as such—and in some cases cannot even be clearly formulated—without the conceptual framework of game theory. Arrow's impossibility theorem, which shows that attempts to combine disparate individual preferences into a "fair" and "democratic" collective choice are foredoomed to failure, was dealt with at length in Chapter 10 and will not be discussed again here. Mackie's discussion of the evolution and stability of moral principles, which also falls into this class, is dealt with below in Section 13.5. A brief summary of the chapter is given in Section 13.6.

13.2 Rationality and Self-Interest

According to Bertrand Russell (1954), " 'Reason' has a perfectly clear and precise meaning. It signifies the choice of the right means to an end that you wish to achieve" (p. 8). This is the *Zweckrationalität* referred to earlier, and it is the interpretation favoured by most contemporary philosophers, but we shall see that its meaning is not as clear and precise as it appears to be. Several philosophers have drawn attention to the way in which the concept of rationality appears to break down in strategic interactions with the underlying strategic structure of the Prisoner's Dilemma game or the N-person Prisoner's Dilemma.

The essential features of the Prisoner's Dilemma game (PDG) were discussed at length in Chapter 6. A brief resumé of the hypothetical problem from which its name derives is as follows. Two people are being held by the police in separate cells, unable to communicate with each other. They are suspected of joint involvement in a serious crime of which they are, in fact, guilty, but the police have insufficient evidence to convict either unless at least one of them discloses certain incriminating evidence. Each prisoner has to choose between concealing the evidence (C) or disclosing it (D). If both conceal, they will be acquitted, and if both disclose, they will be convicted. But if only one of the prisoners discloses the evidence, he will not only be acquitted but will be rewarded for helping the police, while his partner in crime will receive an especially heavy sentence on account of his obstructiveness. The best possible outcome from each prisoner's point of view is acquittal plus reward, the second best is acquittal without reward, the third best is conviction, and the worst is conviction coupled with an especially heavy sentence. These are the prisoners' preferences taking into account all relevant considerations, including their sense of duty to the police, their attitudes towards loyalty and betrayal, and so forth.

The ends that the prisoners wish to achieve are the best possible outcomes from their own points of view, but what are the right means to these ends? Or, to put it differently, what is it rational for each of the prisoners to do?

Each might reason as follows: "If the other prisoner chooses *C*, then the right means to the end that I wish to achieve is to choose *D*, since I shall get the reward by doing so. If, on the other hand, my partner in crime chooses *D*, then the right means to the end is again *D*, since I shall avoid the especially heavy sentence by choosing it. Therefore, although I don't know which course of action he will choose, the rational choice from my own point of view is clearly *D*." But if both prisoners reason in this way they will both be convicted, while if both opt "irrationally" for *C* they will both be acquitted and will achieve their desired ends more effectively. There is evidently a genuine problem of interpreting rationality in this game, which can, of course, crop up in many different contexts in which there is a choice between cooperation (*C*) and defection (*D*) apart from the hypothetical story outlined above.

Some philosophers have offered spurious solutions to the problem. Rescher (1975), for example, discussed the dilemma at considerable length and concluded that "the parties were entrapped in the 'dilemma' because they did not internalize the welfare of their fellows sufficiently. If they do this, and do so in sufficient degree, they can escape the dilemmatic situation" (p. 48). Rescher's argument runs like this:

> The PDG presents a problem for the conventional view of rationality only when we have been dragooned into assuming the stance of the theory of games itself [p. 34]. . . . To *disregard* the interests of others is not rational but inhuman. And there is nothing *irrational* about construing our self-interest in a larger sense that also takes the interests of others into account [p. 39]. . . . Looked upon in its proper perspective, the prisoner's dilemma offers the moral philosopher nothing novel. Its shock-effect for students of political economy inheres solely in their ill-advised approach to *rationality* in terms of prudential pursuit of selfish advantage [p. 40]. . . . When we *internalize* the interests of others, the calculations of self-interest will generally lead to results more closely attuned to the general interests of the group [p. 46, Rescher's italics throughout].

The argument rests on a fundamental misunderstanding of the nature and role of utilities in game theory, and it contains, in addition, an important *non sequitur*. First, the misunderstanding: there is no assumption in game theory that the players' utilities, represented by the payoffs in the matrix, are based on a disregard of each other's interests. On the contrary, the utilities are assumed to reflect their preferences taking all things (including the "welfare of their fellows") into account as their tastes, consciences, and moral principles dictate. The PDG arises from a particular pattern of preferences on the part of the players, and these preferences can arise from altruism or selfishness or any other internally consistent criterion for evaluating the possible outcomes of the game. Rescher treats the payoffs in the PDG matrix as "'raw', first-order utilities" and then proceeds to transform them into "'cooked', other-considering, second-order ones" (p. 46) in order to show how the dilemma can be made to vanish. But there are simply no such things as "raw" utilities in Rescher's sense; they are served ready-cooked in the payoff matrix.

The *non sequitur* is this: accepting for the moment the distinction between "raw" and "cooked" utilities, and even if cooking the utilities in a particular matrix can eliminate the PDG structure (as Rescher demonstrates mathematically), it does not follow that such cooking, if thorough enough, will "generally" allow the players to "escape the dilemmatic situation". It could easily be shown, in fact, that many non-dilemma payoff matrices become PDGs only *after* the utilities have been thoroughly cooked, whereupon the problem of defining rationality is back with a vengeance.

Tuck (1979) attempted to exorcize the N-person Prisoner's Dilemma (NDP), which he calls the "free-rider problem". His solution is also illusory, though for quite different reasons from those which apply to Rescher. The strategic structure of the NPD was examined in detail in Chapter 8; one simple example will suffice for the discussion that follows. During a drought, a person has to decide whether to water his garden or to exercise restraint. Whether all, or some, or none of the other members of the community exercise restraint, the individual (we may assume) is better off, all things considered, watering his garden: this is the "right means to an end" that he wishes to achieve. Restraint on the part of one person is unnecessary if most of the others exercise restraint, and if they do not, it is futile. But if all members of the community reason in this way, each ends up worse off than if all "irrationally" exercise restraint, since the water supply then dries up completely. This is obviously a multi-person generalization of the PDG in which there is once again a cooperative (C) option and a defecting (D) one, and it can arise in a variety of different contexts.

According to Tuck (1979), the dilemma is connected with the ancient *Sorites paradox*. A simple version of the Sorites paradox goes as follows. One stone is clearly not a heap of stones, and the addition of one stone to something that is not a heap can never transform it into a heap; therefore there can never be a heap of stones and it would be futile to try to build one. A similar argument can be used to prove that all men are bald or that all men are hairy. Tuck pinpointed the crux of the NPD in the following version of the Sorites paradox: the water consumption of one person (in the example given earlier) cannot transform a non-critical shortage into a critical one, and, no matter how much water is being used by other members of the community, the consumption of one more person cannot make all the difference. It *seems*, therefore, that the water shortage can never become critical and that there is no point in exercising restraint, just as it *seems* as though a heap of stones can never be built and that it is futile to try to build one. Tuck's conclusion is that the NPD "is not one that should be taken seriously; we have seen that it is connected with a *paradox*, and the essence of paradoxes is that their conclusions should not be believed (for if they were, they would cease to be paradoxes, and become merely good arguments)" (p. 154, Tuck's italics). In Tuck's view, logicians have not yet developed the necessary formal equipment for dealing with the Sorites paradox (and therefore also the NPD), but this is a problem that need not detain the moral philosopher.

There are two comments worth making about this plausible argument. The first is that it seems debatable, to say the least, whether the Sorites is genuinely paradoxical or merely sophistical. A more important point is that the whole argument collapses in the two-person (PDG) version of the dilemma, or for that matter in NPDs with very few players. With just one (or a few) other players, it clearly does (or is quite likely to) make all the difference whether the individual cooperates or defects, but the dilemma is in no way diminished. Even if the Sorites is genuinely paradoxical, therefore, its connection with the NPD is evidently tenuous.

In a thoughtful essay, Martin Hollis (1979) attempted to sharpen and clarify the concept of rationality in the light of the NPD. The major thrust of his argument was that rationality ought to be defined in terms of what is objectively the best way to achieve a certain end, and not in terms of a person's internal beliefs about what is the best way. According to Hollis, a person P acts rationally in choosing a course of action a if and only if the following three conditions are satisfied: (i) of all the possible courses of action available, a is likeliest to realize a goal g; (ii) it is in P's objective interest to realize g; and (iii) Conditions (i) and (ii) are P's reasons for choosing a. It would not be rational for P to choose a merely because of an internal *belief* that a is the best way of achieving g, or because P has an internal *desire* to achieve g; P must choose a for the specified objective reasons. If we do not have external criteria of rationality, then the basis for classifying certain deluded people as mentally ill falls away. It follows from Hollis's argument—though he is unclear on this point—that a person P acts rationally in the NPD if and only if P chooses to defect (to water his garden in the example given earlier) and his reason for doing so is that defecting is in his objective interest. If this conclusion is accepted, however, then a community of rational individuals ends up worse off than a community of irrationalists. Hollis's theory does not come to grips with the central problem of the NPD, the tension between individual and collective rationality.

Bernard Williams (1979) has taken issue with Hollis's theory of rationality and has put forward a contrary point of view. The attempt to define rationality according to external criteria fails, according to Williams, because a person's reasons for performing an action are necessarily internal. If a person believes that the liquid in a certain bottle is gin when in fact it is petrol, and if he wants a gin and tonic, he acts rationally, relative to his false belief, by mixing the liquid with tonic and drinking it, in spite of the fact that it is not in his objective interest to do so. Hollis tried to specify external reasons for actions in his Condition (iii), but in Williams's view, external reason statements, "when definitely isolated as such, are false, incoherent, or really something else misleadingly expressed" (p. 19). Williams concluded that it is not necessarily irrational to cooperate in an NPD. It is rational for a person to cooperate even for purely selfish reasons in some NPDs when "reaching the critical number of those doing C is sensitive to his doing C, or he has reason to think this" (p. 27). A small NPD or a PDG would presumably meet this requirement. But, according to Williams, it can be rational to choose cooperatively even if this requirement is not met, and it is rational for a society to educate people to have cooperative motivations in strategic interactions of the NPD type.

The most thoroughgoing analyses of rational action in the light of the PDG and NPD are those of Derek Parfit (1979a, 1979b). Parfit came to the same conclusion as Williams, that cooperation can be a rational course of action. He put his finger on the essence of the problem by pointing out that "common-sense morality", which involves performing those acts that best achieve the ends we feel we ought to achieve, is "self-defeating" in these games. If all rather than none successfully follow common-sense morality, their ends are worse achieved than if none do so. Can common-sense morality be defended in the light of this? Parfit's answer is that "this depends on our view about the nature of morality. On most views, the answer is 'No'. But I must here leave the question open" (1979b, p. 564).

Various ways out of the dilemma were outlined by Parfit. One class of solutions involves changing the rules of the game or the players' payoff functions, by making it impossible or unprofitable for anyone to defect, for example. Heavy fines for watering one's garden during a drought, coupled with intensive policing, could change the strategic structure of the underlying game so that it ceased to be an NPD. These are "political" solutions, and in Parfit's view they are often undesirable. Sometimes they are infeasible: in the case of the nuclear arms race there is no government with the power to change the rules or the payoff structure of the game or to enforce cooperation. Whether political solutions are as undesirable as they seem will be discussed further in connection with Rousseau's social contract theory in Section 13.4. But Parfit argued that it is better, where possible, to find ways of making people want to cooperate for moral reasons in spite of the existence of the dilemma and an absence of coercion; these he called "moral" solutions. How, then, are we to achieve moral solutions? "Prisoner's Dilemmas need to be explained. So do their moral solutions. Both have been too little understood" (1979b, p. 544).

13.3 Kant's Categorical Imperative

A persuasive moral solution to the PDG and NPD can be derived from Immanuel Kant's *categorical imperative*, and Kant's views can be greatly clarified by applying them to these and other games. That the categorical imperative is in need of clarification is shown by the seemingly endless debates among moral philosophers about its precise meaning (see, for example, the critical essays reprinted in Wolff, 1968, pp. 211–336).

Kant formulated the categorical imperative in various different ways, but his first formulation, which is the one most often quoted, is this: "Act only on such a maxim through which you can at the same time will that it should become a universal law" (Kant, 1785, p. 52). A maxim is a purely personal rule of conduct which is categorical (unconditional) if it takes the form "Always do *a*"; and a universal law is a categorical rule prescribing the conduct of all people. The categorical imperative is intended as a fundamental principle of morality.

In his earlier *Critique of Pure Reason*, Kant scorned examples as the "go-carts of philosophy", but in his *Grundlegung* and his *Critique of Practical Reason* he supplied a handful of examples to illustrate the categorical imperative. His least troublesome example is the following (freely translated) maxim: "Always borrow money when in need and promise to pay it back without any intention of keeping the promise" (1785, p. 54). One cannot will this maxim to be a universal law because it cannot be universalized without generating a contradiction. If everybody habitually broke promises, there could be no promises because nobody would believe them. People could utter the words "I promise" but they could express only an intention, not a true promise, because a promise needs a promisee as well as a promisor, and there would be no promisees. To will the maxim to be a universal law thus leads to a logical contradiction: if everybody habitually breaks promises then nobody can do so because promises will have ceased to exist.

All of Kant's other examples are considered by many moral philosophers to be obscure. Is it morally acceptable, for example, to adopt the maxim of always refusing to help others who are in need or distress, provided one does not (inconsistently) demand help from others? A state of affairs in which this is a universal law is certainly possible in theory—and it is not difficult to imagine—but, according to Kant, it is impossible for a person to *will* this maxim to be a universal law, because "a will which decided in this way would be at variance with itself. . . . By such a law of nature sprung from his own will, he would rob himself of all hope of the help he wants for himself" (1785, p. 56). Kant was clearly referring to some kind of inconsistency, but what exactly he meant by "a will at variance with itself" is not immediately obvious. Why does a person's will become inconsistent when universalized in cases of this type?

When formulated in game theory terms, Kant's remarks become perfectly clear. The players in a game are, by definition, motivated to achieve what each considers to be the best possible outcome. If the game is symmetric, that is to say if it presents the same choice to every player, then the categorical imperative can be reformulated as follows: "Choose only a strategy which, if you could will it to the chosen by all of the players, would yield a better outcome from your point of view than any other". In a pure coordination game (see Chapter 3), any strategy is morally acceptable according to this criterion. In a strictly competitive game (see Chapter 4), only a minimax strategy passes the test. In a PDG or an NPD only a cooperative choice is morally acceptable for the following reason: the alternative defecting choice, if universalized, is at variance with a player's own interest. In other words, although it is better from the perspective of each individual player to defect (for example by making false promises) whether or not the other player(s) cooperate, it would be at variance with each individual's motive to will that every player should defect because each considers a universally cooperative outcome to be better than a universally defecting one. This nicely captures the inconsistency that Kant probably had in mind, and incidentally shows how difficult it is to formulate essentially strategic problems without the conceptual framework of game theory, as Kant was forced to do.

An unexpected and neglected feature of the categorical imperative is that it prescribes selfish rather than altruistic behaviour in some types of strategic interactions. An unambiguous example is the Battle of the Sexes game (see Section 6.4): A man and a woman have to choose between two options for an evening's entertainment. The man prefers one of the options and the woman prefers the other, but each would rather go out together than alone. If both opt altruistically for their less preferred alternatives, then each ends up attending a disliked entertainment alone; but if both choose selfishly, then each has a somewhat more enjoyable evening out. According to the categorical imperative, therefore, each ought to adopt a selfish maxim in the Battle of the Sexes; an altruistic maxim, if universalized, would be at variance with his or her will. But this example also reveals a limitation of the categorical imperative, since the outcome would be better still, from both players' viewpoints, if the man acted on an altruistic maxim and the woman on a selfish one or vice versa, enabling them to spend the evening together rather than alone. Is it morally unacceptable for the man or the woman to act on an altruistic maxim and at the same time will that the other person should choose selfishly? The categorical imperative seems morally persuasive only in games with symmetric equilibrium points.

Whether it is reasonable to expect that people would ever adopt the categorical imperative as a moral principle is another matter altogether. The sections that follow throw some light on questions of that type.

13.4   Rousseau's Social Contract

Jean-Jacques Rousseau's *The Social Contract*, first published in 1762, is one of the most influential books in the history of political and social philosophy, although some of the pivotal ideas in it are generally considered to be opaque or inconsistent. Many of the ideas underlying the social contract theory were expounded earlier by Thomas Hobbes, especially in Chapters 13 to 17 of *Leviathan* (first published in 1651), and some of the comments in this section are as pertinent to Hobbes's version of the theory as to Rousseau's. The "fundamental problem to which the social contract provides a solution", according to Rousseau, is "to find a form of association that defends and protects, with the whole common force, the person and goods of each member, and in which each, by uniting with all, obeys only himself and remains as free as before" (Bk. I, chap. vi).

A considerable amount of confusion has arisen from Rousseau's emphatic distinction between the *general will*, which the social contract is intended to promote, and the *will of all*, which he says is frequently in opposition to it: "There is often a great difference between the will of all and the general will; the latter regards only the common interest; the former regards private interests, and is merely the sum of particular desires" (Bk. II, chap. iii). Commentators have had difficulty explaining how the common interest can be anything other than the sum of private interests. As Runciman and Sen (1965) were apparently the first to point out, the distinction becomes transparent when couched in the framework of game theory. In an NPD, for example, it is in the "private interest" of each player to defect, but it is in their "common interest" to cooperate. If the players have regard only to their private interests, then the result is joint defection, which reflects "merely the sum of particular desires". But if they form a social contract through which joint cooperation can be ensured, then "each, by uniting with all" enjoys a more favourable outcome.

This is not as straightforward as it seems, however. If a purely voluntary agreement is reached, then it may still be in the interest of an individual to defect provided that the act of breaking the agreement is not too disagreeable; a person may prefer the outcome of defecting irrespective of how few or how many of the others cooperate. In other words, the voluntary agreement may leave the underlying NPD strategic structure intact. And if all players pursue their private interests by defecting in spite of the agreement, the agreement will collapse.

It follows from this that the players may be willing to enter into a *binding* contract in which joint cooperation is enforced by coercive means. Even if most people are disposed to defect from a voluntary agreement by watering their gardens during a drought, say, they may appreciate the fact that each of them would be better off under a system of enforced cooperation, since universal cooperation is better for each person than universal defection. It is presumably this idea which lies behind Rousseau's seemingly paradoxical remarks about enforcement:

> Man is born free, and everywhere he is in chains [Bk I, chap. i]. . . . In order that the social contract should not be a vain formula, it tacitly includes an undertaking, which alone can give force to the others, that whosoever refuses to obey the general will shall be constrained by the whole body: this means nothing other than that one forces him to be free [Bk. I, chap. vii]. . . . The undertakings which bind us to the social body are obligatory only because they are mutual, and their nature is such that in fulfilling them we cannot work for others without working also for ourselves [Bk. II, chap. iv].

Interpreted in the light of the NPD, Rousseau's remarks about enforcement become perfectly lucid. A binding agreement among the players to cooperate, with coercive means of enforcement, seems the only way to protect the interests of all players when moral solutions cannot be relied upon, that is to say when all or many would otherwise be disposed to defect. The players would actively desire to be constrained in this way if they were rational, because it guarantees a better outcome for everyone. People can improve the quality of their lives by freely choosing to put themselves in chains. "Mutual coercion mutually agreed upon" (Hardin, 1968, p. 1247) is in the rational self-interests of the participants in strategic interactions of the NPD type, which are extremely common in everyday life (see Chapter 8). Enforced cooperation improves the payoffs to most of the players and to their fellows; by fulfilling their obligations they serve their own interets as well as the interests of others.

One of the most perplexing aspects of Rousseau's views on the social contract is his opposition to political parties and other interest groups within the community: "It is important, therefore, in order to have a good expression of the general will, that there should be no partial association within the state, and that each citizen should think of himself alone" (Bk. II, chap. iii). Farquharson (1969, pp. 77–80) provided a rationale, based on the theory of coalition formation in cooperative games and strategic voting, by which Rousseau's hostility to political parties can be explained. A much simpler rationale than Farquharson's can now be given in terms of the Dollar Auction game.

The rules of the Dollar Auction game, discussed in detail in Section 8.6, are (briefly) as follows. Several players bid for a dollar bill according to the usual conventions of auctions, except that the highest bidder *and* the second-highest bidder have to pay the auctioneer amounts corresponding to their last bids and only the highest bidder receives the dollar bill in return. In a "state of nature", as Rousseau calls it, that is to say when there is no social contract, it is impossible for any player to enter the bidding without running the risk of substantial loss. But, by popular vote, the players might adopt a social contract, according to which one player, nominated in advance, bids 1 cent while the others are forbidden to bid, and the resulting 99-cent prize is divided equally among all the players. This reflects the general will in so far as there is no other outcome that is better for everyone, so the social contract would be supported by all rational players. (In conventional auctions in Britain, coordinated bidding by prior arrangement is illegal in terms of the Auction Bidding Agreements Act of 1937, precisely because it gives the bidders a big advantage.)

Suppose, however, that the players split into political parties or factions, each proposing to modify the social contract to enable their own members to distribute the lion's share of the 99-cent prize among themselves alone. If the various proposals were put to a popular vote, then one of the parties would gain a majority and be able to impose its will on the others. Its members would clearly benefit from this change, and the excluded players, being bound by the social contract, would be powerless to do anything about it; but it would evidently not reflect the general will. In Rousseau's words, when "partial associations form at the expense of the whole, the will of each of these associations becomes general in relation to its members, and particular in relation to the State" (Bk. II, chap. iii). The sense in which the social contract may be said to promote the general will, and partial associations to undermine it, is given a precise interpretation in this example, and the apparent inconsistencies in Rousseau's thinking are reconciled.

13.5 Evolution and Stability of Moral Principles

In two important papers, Mackie (1978, 1982) showed how the theory of evolutionary games (discussed in depth in Chapter 12) can be used (a) to account for the nature and evolution of existing moral principles, and (b) to indicate constraints which must be taken into account in proposing new moral principles if these proposals are to be workable in practice.

Moral principles are cultural rather than physical traits, and they are transmitted through human communication rather than through genetic inheritance. But there are some cultural traits that share certain logical properties in common with genes and are therefore subject to evolution through a kind of natural selection. This idea can be traced back to Karl Popper (1961), but its best known proponent is Richard Dawkins (1976). Dawkins called cultural traits that are subject to natural selection *memes*—a word designed to resemble "genes"—because they are sustained by memory and mimicry. Apart from moral principles, typical examples of memes are tunes, fashions, traditions, and theories.

Genes and memes are alike in the following respects. First, they are self-replicators: in suitable environments they produce multiple copies of themselves; but, secondly, in the process of replication errors occur, and the new forms (mutants) also produce copies of themselves. Finally, a new form may be fitter or less fit than an established one, as measured by the number of copies that it produces, and only the fittest survive in the struggle for existence. Cultural evolution can, of course, proceed much more rapidly than biological evolution because it is not limited by the reproduction rate of the species.

A meme will spread through a population rapidly if there is something about it that makes it better able than the available alternatives to infect people's minds, just as germs spread when they are able to infect people's bodies. This analogy draws attention to the fact that the fittest memes are not necessarily ones that are beneficial to society as a whole. Functionalist theories in sociology and social anthropology, in contrast, are based on the assumption that memes which are beneficial to the society are the only ones that can survive.

One of Mackie's models can be summarized roughly as follows. Consider the altruistic moral principle of always helping other people who are in need or distress; this principle was discussed earlier in connection with Kant's categorical imperative. We may assume that the costs associated with helping others—in terms of time and energy expenditure, inconvenience, and so forth—outweigh any direct and immediate benefits, such as feelings of satisfaction for having acted altruistically, but that these costs are generally less than the benefits to the person being helped. A third necessary assumption is that people pursue individual rationality by choosing the best means to the ends they wish to achieve. Now in a society in which indiscriminate altruism was the ethical norm, an individual who adopted a purely selfish principle of accepting help but never offering it would receive a higher expected payoff, enjoying all the benefits but incurring none of the costs of indiscriminate altruism. This new selfish morality would therefore tend to be imitated by others and would spread until it had infected the whole population. In the terminology of evolutionary games, the selfish morality would be an evolutionarily stable strategy (ESS).

Intermediate between indiscriminate altruism and selfishness is the moral principle of reciprocal altruism: reciprocal altruists are willing to help only those who have not acted in a selfish manner towards them in the past. In a society consisting initially of selfish individuals only, if the new moral principle of reciprocal altruism arose, it would spread through the population and wipe out the selfish morality. The expected payoffs to the reciprocal altruists would be higher because they would fare no worse in interactions with the selfish individuals but would do better than selfish individuals in interactions with their own kind. Natural selection therefore favours an ESS in which reciprocal altruism is the norm, and this, according to Mackie, is the moral principle that is in fact adopted by most people.

Christian teaching, of course, advocates indiscriminate altruism. It is therefore interesting to investigate the logical consequences, according to Mackie's model, of indiscriminate altruism in a mixed population of reciprocal altruists and selfish individuals, a population perhaps in the process of evolving to a pure ESS of reciprocal altruism. It turns out that the introduction of indiscriminate altruists into this population would tend to shift the direction of evolution towards selfishness. This is because selfish individuals would get more out of interactions with the indiscriminate altruists than reciprocal altruists would; they would enjoy the benefits without incurring the costs. In other words, the presence of indiscriminate altruists endangers the healthy reciprocal altruist morality by enabling selfishness to prosper, and if the indiscriminate altruists were sufficiently numerous, the selfish strategy would be an ESS and both kinds of altruists would be wiped out. Selfishness is a relatively ineffective strategy in a population of reciprocal altruists, but it pays better than reciprocal altruism when there are enough indiscriminately altruistic "suckers" to take advantage of. According to Mackie (1978):

> this seems to provide fresh support for Nietzsche's view of the deplorable influence of moralities of the Christian type. But in practice there may be little danger. After two thousand years of contrary moral teaching, reciprocal altruism is still dominant in all human societies. . . . Saintliness is an effective topic for preaching, but with little practical persuasive force (p. 464).

The evolutionary consequences of Christian morality in Mackie's model provide a vivid illustration of what Popper (1969) described as "the *main task of the theoretical social sciences. It is to trace the unintended social repercussions of intentional human actions*" (p. 342, Popper's italics).

13.6 Summary

At the beginning of the chapter, several reasons were put forward to account for the fact that formal game theory seldom provides straightforward solutions to practical problems of strategy. Human beings generally handle complex strategic problems by satisficing, that is to say by searching just long enough to find strategies that are satisfactory or that suffice. Strategic choices in everyday life, since they affect other people, often raise moral problems. There are important limitations to what can be said about moral problems within the conceptual framework of game theory, but the theory can throw light on existing problems in moral philosophy, and it can also draw attention to previously unrecognized problems.

In Section 13.2, the strategic implications of rationality and the pursuit of self-interest were examined. The concept of rationality, as it is normally interpreted by contemporary philosophers, appears to break down in the Prisoner's Dilemma game and the N-person Prisoner's Dilemma. Several attempts by philosophers to come to grips with this and other related problems were critically reviewed. Section 13.3 centred on Kant's categorical imperative as a principle of morality. Game theory elucidates Kant's essentially strategic ideas and also reveals some limitations of his principle. Rousseau's ideas about the social contract, discussed in Section 13.4, are often considered to be opaque and inconsistent in parts, particularly in the distinction between the general will and the will of all and in the objection to political parties. Reformulated in game theory terms, however, both aspects of the theory become clear and the apparent inconsistencies disappear. In Section 13.5, the use of evolutionary games to explain the nature and development of existing moralities and to indicate certain constraints on any new moralities which might be proposed was outlined. The application of game theory to these questions leads to conclusions that are sometimes quite unexpected.

APPENDIX A

A Simple Proof of the Minimax Theorem

A.1 Introductory Remarks

Books on non-mathematical aspects of game theory traditionally extol the minimax theorem and make extensive use of it without offering any arguments to persuade the reader of its truth. As a consumer of such books, I have always felt cheated by this, and it seems reasonable to suppose that others have reacted similarly. Since there seems to be no source to which mathematically unsophisticated readers can turn in order to understand the theorem, I have attempted in the pages that follow to provide a simple, self-contained proof with each step spelt out as clearly as possible both in symbols and in words. It was worked out in collaboration with my mathematical colleague Roy Davies, but is really a variation on a well-known theme: see, for example, Blackwell and Girshick (1954) for a more advanced version. Non-mathematicians will find the argument much easier to follow than the standard presentations in mathematical textbooks; but elementary algebra and geometry are assumed, and a complete understanding demands some knowledge of the rudiments of probability, which are explained from first principles in—for example—Colman (1981, chap. 4).

A.2 Preliminary Formalization

A finite, two-person, zero-sum game is specified by a rectangular array of numbers (a_{ij}) having m rows and n columns, commonly called a *payoff matrix*. The numbers are the payoffs to Player I; since the game is zero-sum, Player II's payoffs are simply the negatives of these numbers.

According to the rules of the game, Player I chooses a strategy corresponding to one of the rows, and simultaneously—or, what amounts to the same thing, in ignorance of I's choice—II chooses a strategy corresponding to one of the columns. The number at the intersection of the chosen row and column is the corresponding payoff. Thus if Player I chooses row i and Player II chooses column j, the number a_{ij} at the intersection is the amount gained by I and lost by II; in other words, the amount a_{ij} is transferred from II to I. Complete information is assumed: both players know the payoff matrix and each knows that the other is in possession of complete information.

Instead of deliberately selecting a *pure* strategy—a specific row or column—a player may use a randomizing device to choose among them. A player who has two pure strategies, for example, may decide to allow his or her choice to depend on the toss of an unbiased coin. A player who chooses in this way is said to be using a *mixed* strategy. In general, a mixed strategy assigns a predetermined probability to each available pure strategy; in the example above, the probabilities are 1/2 and 1/2. A mixed strategy can thus be represented by a string of non-negative numbers of length m (for Player I) or n (for Player II) that sums to 1. A mixed strategy for Player I can accordingly be written

$$(x) = (x_1, \ldots, x_m),$$

and a mixed strategy for Player II

$$(y) = (y_1, \ldots, y_n),$$

where $x_1 \geq 0, \ldots, x_m \geq 0, y_1 \geq 0, \ldots, y_n \geq 0, x_1 + \cdots + x_m = 1$, and $y_1 + \cdots + y_n = 1$. A pure strategy can be viewed as a special case of a mixed strategy in which a probability of 1 is assigned to one of the x_i or y_j and 0 to each of the others.

If Player I uses a mixed strategy (x) and II uses a mixed strategy (y), then row i will be chosen with probability x_i and column j with probability y_j. Since these events are independent, the payoff a_{ij} will occur with probability $x_i y_j$. The *expected payoff* is then simply a weighted average of all the payoffs a_{ij}, each one occurring with probability $x_i y_j$, and it can be written $\sum a_{ij} x_i y_j$, where $i = 1, \ldots, m$ and $j = 1, \ldots, n$.

A.3   The Minimax Theorem

Since the expected payoff represents I's average gain and II's average loss, I wants to maximize it and II wants to minimize it. If Player I knew in advance that II was going to use a specific mixed strategy (y'), then I's best counter-strategy would be one which maximizes the expected payoff against (y'); the expected payoff would then be

$$\max_{(x)} \sum a_{ij} x_i y'_j.$$

Similarly, if II knew I's mixed strategy (x') in advance, II could use a counter-strategy which minimizes the expected payoff, yielding

$$\min_{(y)} \sum a_{ij} x'_i y_j.$$

These counter-strategies cannot be used in practice because neither player has foreknowledge of the other's intentions. Player I can nonetheless maximize his or her *security level* by *assuming* pessimistically that any strategy (x) will be met by II's corresponding minimizing counter-strategy, and by choosing (x) so as to maximize the expected payoff under this assumption. Player I thus ensures that the expected payoff will be no less than

$$\max_{(x)} \min_{(y)} \sum a_{ij} x_i y_j.$$

A similar argument shows that II's security level is maximized by the use of a strategy (y) which minimizes the expected payoff on the assumption that Player I will reply to any (y) with the corresponding maximizing counter-strategy; the expected payoff will then be no more than

$$\min_{(y)} \max_{(x)} \sum a_{ij} x_i y_j.$$

THEOREM. *If* (a_{ij}) *is any m by n payoff matrix, then*

$$\max_{(x)} \min_{(y)} \sum a_{ij} x_i y_j = \min_{(y)} \max_{(x)} \sum a_{ij} x_i y_j, \tag{1}$$

where $(x) = (x_1, \ldots, x_m)$ *and* $(y) = (y_1, \ldots, y_n)$ *represent all strings of non-negative numbers of length m, n and sum 1.*

A.4   Proof

A.4.1.   Denote the left-hand side of equation (1) by v and the right-hand side by w. Assume that the payoff matrix (a_{ij}) is given, and let $(x') = (x'_1, \ldots, x'_m)$ be that (x) for which the left-hand maximum is attained and $(y') = (y'_1, \ldots, y'_n)$ that (y) for which the right-hand minimum is attained. Since the minimum of a variable quantity is less than or equal to any particular value of it, and the opposite holds for a maximum,

$$v = \min_{(y)} \sum a_{ij} x'_i y_j \leqslant \sum a_{ij} x'_i y'_j \leqslant \max_{(x)} \sum a_{ij} x_i y'_j = w. \tag{2}$$

We have proved that $v \leqslant w$, and it will now suffice to establish that $v \geqslant w$. First, after providing a geometrical model of a game in A.4.2, we shall prove in A.4.3 to A.4.6 that if $w > 0$, then $v > 0$.

A.4.2.   If Player I chooses row 1 and II uses a mixed strategy (y), then the expected payoff will be the following weighted average of the n numbers in row 1:

$$a_{11} y_1 + \cdots + a_{1n} y_n = \sum a_{1j} y_j.$$

The same mixed strategy (y) in combination with Player I's row 2 would yield an expected payoff of $\sum a_{2j} y_j$, and so on for each of the m rows. Any of Player II's mixed strategies (y) can therefore be represented by a point in m-dimensional space, each coordinate of the point corresponding to the expected payoff of (y) in combination with one of Player I's pure strategies (rows). The coordinates of the point determined by (y) are then

$$(\textstyle\sum a_{1j}y_j , \ldots, \sum a_{mj}y_j).$$

Let C be the set of all these points, where as usual $(y) = (y_1, \ldots, y_n)$ represents all strings of non-negative numbers satisfying $y_1 + \cdots + y_n = 1$.

If $m = 2$, C is two-dimensional and can be shown pictorially; if $m = 3$, it is three-dimensional. For $m > 3$, C does not correspond to any object in ordinary space, but it can be handled algebraically by generalizing straight-forwardly from the two- and three-dimensional cases. The game shown in Matrix A.1 with $m = 2$ will suffice to give a geometrical interpretation to C. The representation of this game in two-dimensional space is shown in Figure A.1.

Matrix A.1

II

		1	2	3
I	1	2	5	1
	2	3	1	−1

Every point in C corresponds to a strategy for Player II. The first coordinate of the point is the payoff if Player I simultaneously chooses row 1, and the second coordinate is the payoff if Player I chooses row 2. The vertices of C correspond to II's pure strategies, and all other points on the boundary or inside C are weighted averages of its vertices and correspond to II's mixed strategies. As an illustration, the point $(3, 1)$ in C corresponds to $(y) = (2/7, 3/7, 2/7)$, since the expected payoff against row 1 is then

$$\textstyle\sum a_{1j}y_j = (2)(2/7) + (5)(3/7) + (1)(2/7) = 3,$$

and against row 2 the expected payoff is

$$\textstyle\sum a_{2j}y_j = (3)(2/7) + (1)(3/7) + (-1)(2/7) = 1,$$

so the coordinates of the corresponding point are $(3, 1)$.

If Player I uses a mixed strategy (x), the expected payoff will be a weighted average of the coordinates of the point in C corresponding to II's strategy. Thus if Player I uses (x) and Player II uses (y), the expected payoff is

$$x_1(\textstyle\sum a_{1j}y_j) + \cdots + x_m(\sum a_{mj}y_j) = \sum a_{ij}x_iy_j.$$

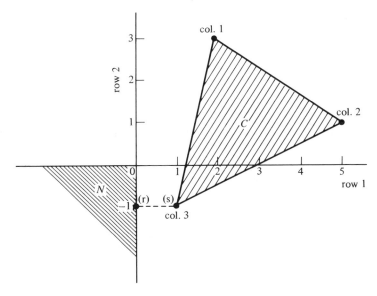

FIG. A.1   Geometric interpretation of the game described in the text, showing C, N, and the line segment between (s) and (r)—shown dashed.

A.4.3.   Now suppose that $w > 0$. According to the definition of w, for any (y') there exists (x) such that

$$\sum a_{ij} x_i y'_j > 0,$$

which means that for any point in C, at least one of its coordinates must be positive. The point cannot, therefore, lie in the region N consisting of points with all coordinates negative or zero. We have established that C and N have no common point.

A.4.4.   Let $(r) = (r_1, \ldots, r_m)$ be the point of N nearest to C, and let $(s) = (s_1, \ldots, s_m)$ be the point of C nearest to (r). Now if r_1 is replaced by any negative or zero number r'_1, the point $(r'_1, r_2, \ldots, r_m)$ is also in N, and it is therefore no nearer than (r) to (s). This is easy to visualize in two or three dimensions (see Fig. A.1); in m dimensions

$$(s_1 - r'_1)^2 + (s_2 - r_2)^2 + \cdots + (s_m - r_m)^2$$
$$\geq (s_1 - r_1)^2 + (s_2 - r_2)^2 + \cdots + (s_m - r_m)^2.$$

Generalizing from the three-dimensional case, the repeated application of Pythagoras' theorem shows that the left-hand side of this inequality is the square of the distance between (s) and $(r'_1, r_2, \ldots, r_m)$ and the right-hand side is the square of the distance between (s) and (r)—this is a well-known result in elementary coordinate geometry or vector algebra. The inequality simplifies to

$$(s_1 - r_1')^2 \geq (s_1 - r_1)^2. \tag{3}$$

Now if $s_1 \leq 0$, then a possible value of r_1' is $r_1' = s_1$, in which case the left-hand side of (3) is zero and therefore $r_1 = s_1$. If, on the other hand, $s_1 > 0$ and we take $r_1' = 0$, then (3) becomes $s_1^2 \geq (s_1 - r_1)^2$, which simplifies to $2s_1 r_1 \geq r_1^2$. In that case, since $r_1 \leq 0$ and s_1 is a positive number, the inequality is satisfied only if $r_1 = 0$. It follows from these two results that $s_1 \geq r_1$ and, since either $r_1 = s_1$ or $r_1 = 0$, that $(s_1 - r_1)r_1 = 0$. The same argument can be used to show that $s_2 \geq r_2$ and $(s_2 - r_2)r_2 = 0$, and so on. Thus

$$s_1 \geq r_1, \; s_2 \geq r_2, \ldots, s_m \geq r_m,$$

and

$$(s_1 - r_1)r_1 + (s_2 - r_2)r_2 + \cdots + (s_m - r_m)r_m = 0.$$

A.4.5.   For any number t between 0 and 1 and any point (a) in C, the point

$$t(a) + (1 - t)(s) = (ta_1 + (1 - t)s_1, \ldots, ta_m + (1 - t)s_m)$$

on the straight line between (a) and (s) is a weighted average of the points (a) and (s). Since (a) and (s) belong to C, they in turn are weighted averages of the vertices of C, and therefore so is the point $t(a) + (1 - t)(s)$. This means that the point $t(a) + (1 - t)(s)$ also belongs to C, and it is thus no nearer than (s) to (r):

$$(ta_1 + (1 - t)s_1 - r_1)^2 + \cdots + (ta_m + (1 - t)s_m - r_m)^2$$
$$\geq (s_1 - r_1)^2 + \cdots + (s_m - r_m)^2.$$

This simplifies to

$$2(a_1 - s_1)(s_1 - r_1) + \cdots + 2(a_m - s_m)(s_m - r_m)$$
$$\geq -t\{(a_1 - s_1)^2 + \cdots + (a_m - s_m)^2\}.$$

Since t can be arbitrarily close to zero, it follows that

$$(a_1 - s_1)(s_1 - r_1) + \cdots + (a_m - s_m)(s_m - r_m) \geq 0.$$

This can be rewritten

$$a_1(s_1 - r_1) + \cdots + a_m(s_m - r_m)$$
$$\geq \{(s_1 - r_1)^2 + \cdots + (s_m - r_m)^2\} + \{(s_1 - r_1)r_1 + \cdots + (s_m - r_m)r_m\}.$$

In view of the result in A.4.5, we may conclude that

$$a_1(s_1 - r_1) + \cdots + a_m(s_m - r_m) \geq (s_1 - r_1)^2 + \cdots + (s_m - r_m)^2.$$

The right-hand side of this inequality is a sum of squares and cannot, therefore, be negative, hence

$$a_1(s_1 - r_1) + \cdots + a_m(s_m - r_m) \geq 0. \tag{4}$$

A.4.6.   Each of the numbers $s_1 - r_1, \ldots, s_m - r_m$ is positive or zero according to A.4.4, and not all are zero because (s) and (r) are different points, so their sum q is positive. Let $x_1^0 = (s_1 - r_1)/q, \ldots, x_m^0 = (s_m - r_m)/q$. Then x_1^0, $\ldots, x_m^0$ are the results of dividing positive or zero numbers by a positive number. It follows that $x_1^0, \ldots, x_m^0$ are each non-negative, and it is clear that $x_1^0 + \cdots + x_m^0 = 1$ since $(s_1 - r_1) + \cdots + (s_m - r_m) = q$ and $q/q = 1$. Thus $(x^0) = (x_1^0, \ldots, x_m^0)$ satisfies the requirements of a mixed strategy for Player I.

Dividing each term of (4) by the positive quantity q, and bearing in mind that $(s) \neq (r)$,

$$a_1 x_1^0 + \cdots + a_m x_m^0 > 0$$

for every point (a) in C. According to the definition of C in A.4.2, the coordinates of (a) are $(\Sigma a_{1j} y_j, \ldots, \Sigma a_{mj} y_j)$. We have therefore found a mixed strategy (x^0) for Player I such that

$$x_1^0 (\Sigma a_{1j} y_j) + \cdots + x_m^0 (\Sigma a_{mj} y_j) > 0$$

for every (y), so that

$$\min_{(y)} \Sigma a_{ij} x_i^0 y_j > 0.$$

Since this holds for (x^0), it must hold for the mixed strategy (x) that maximizes

$$\min_{(y)} \Sigma a_{ij} x_i y_j.$$

This means that

$$\max_{(x)} \min_{(y)} \Sigma a_{ij} x_i y_j > 0.$$

This completes the proof that if $w > 0$, then $v > 0$.

A.4.7.   Now let k be any number, and consider the m by n payoff matrix which has $a_{ij} - k$ in place of a_{ij} for every i and j. All payoffs for pure strategies in the corresponding game are evidently reduced by k, and therefore so are those for mixed strategies, so that v and w are replaced by $v - k$ and $w - k$ respectively. By what we have just proved, if $w - k > 0$, then $v - k > 0$. This implies that if $w > k$, then $v > k$. Since k can be arbitrarily close to w, it follows that $v \geqslant w$. But we have seen in A.4.1 that $v \leqslant w$. It follows that $v = w$, that is,

$$\max_{(x)} \min_{(y)} \sum a_{ij}x_i y_j = \min_{(y)} \max_{(x)} \sum a_{ij}x_i y_j.$$

This completes the proof of the minimax theorem.

REMARK. Since the outer two numbers in Equation (2) are equal, their common value is also equal to the term in the middle. This implies that (x') and (y') are mixed strategies yielding the same security level to the players using them. In other words, mixed strategies yielding the same security level to each player are bound to exist in any finite, two-person, zero-sum game; that is, any such game has an equilibrium point in mixed strategies.

References

Alcock, J. E. and Mansell, D. (1977). Predisposition and behaviour in a collective dilemma. *Journal of Conflict Resolution*, 21, 443–57.

Apfelbaum, E. (1974). On conflicts and bargaining. *Advances in Experimental Social Psychology*, 7, 103–56.

Argyle, M. (1969). *Social Interaction*. London: Methuen.

Arrow, K. J. (1951). *Social Choice and Individual Values*. New York: Wiley.

Arrow, K. J. (1963). *Social Choice and Individual Values*. 2nd ed. New York: Wiley.

Atkinson, R. C. and Suppes, P. (1958). An analysis of two-person game situations in terms of statistical learning theory. *Journal of Experimental Psychology*, 55, 369–78.

Barbera, S. (1977). Manipulation of social decision functions. *Journal of Economic Theory*, 15, 266–78.

Becker, G. M. and McClintock, C. G. (1967). Value: behavioural decision theory. *Annual Review of Psychology*, 18, 239–86.

Bernoulli, D. (1738). Specimen theoriae novae de mensura sortis. In *Comentarii Academii Scientarum Imperialis Petropolitanae*, 5, 175–92. Translated by L. Sommer in *Econometrica* (1954), 22, 23–6.

Bernstein, I. S. and Gordon, T. P. (1974). The function of aggression in primate societies. *American Scientist*, 62, 304–11.

Bixenstine, V. E. and O'Reilly, E. F. Jr. (1966). Money vs. electric shock as payoff in a Prisoner's Dilemma game. *Psychological Record*, 16, 251–64.

Bjurulf, B. H. and Niemi, R. G. (1981). Order-of-voting effects. In M. J. Holler (Ed.), *Power, Voting, and Voting Power*. Würzburg: Physica-Verlag.

Black, D. (1948a). The decisions of a committee using a special majority. *Econometrica*, 16, 245–61.

Black, D. (1948b). On the rationale of group decision making. *Journal of Political Economy*, 56, 23–34.

Black, D. (1958). *The Theory of Committees and Elections*. Cambridge: Cambridge University Press.

Blackwell, D. and Girshick, M. A. (1954). *Theory of Games and Statistical Decisions*. New York: Wiley.

Blair, D. H., Bordes, G., Kelly, J. S., and Suzumura, K. (1976). Impossibility theorems without collective rationality. *Journal of Economic Theory*, 13, 361–79.

Blau, J. H. (1957). The existence of a social welfare function. *Econometrica*, 25, 302–13.

Blau, J. H. (1972). A direct proof of Arrow's theorem. *Econometrica*, 40, 61–7.

Blin, J.-M. (1973). Intransitive social orderings and the probability of the Condorcet effect. *Kyklos*, 26, 25–35.

Bonacich, P., Shure, G. H., Kahan, J. P., and Meeker, R. J. (1976). Cooperation and group size in the *N*-Person Prisoners' Dilemma. *Journal of Conflict Resolution*, 20, 687–706.

Borda, J.-C. de (1781). Mémoire sur les élections au scrutin. *Mémoires de l'Académie Royale des Sciences*, 657–665. Trans. A. de Grazia, *Isis* (1953), 44, 42–51.

Boring, E. G. (1950). *A History of Experimental Psychology*, 2nd ed. New York: Appleton-Century-Crofts.

Bowen, B. D. (1972). Occurrence of the paradox of voting in U.S. senate roll call votes. In R. G. Niemi and H. F. Weisberg (Eds.), *Probability Models of Collective Decision Making*, pp. 181–203. Columbus: Charles E. Merrill.

Braiker, H. B. and Kelley, H. H. (1979). Conflict in the development of close relationships. In

R. L. Burgess and T. L. Huston (Eds), *Social Exchange in Developing Relationships*, pp. 135–68. New York: Academic Press.

Braithwaite, R. B. (1955). *Theory of Games as a Tool for the Moral Philosopher*. Cambridge: Cambridge University Press.

Brams, S. J. (1975). *Game Theory and Politics*. New York: Free Press.

Brams, S. J. (1976). *Paradoxes in Politics: An Introduction to the Nonobvious in Political Science*. New York: Free Press.

Brams, S. J. (1977). Deception in 2 × 2 games. *Journal of Peace Science*, **2**, 171–203.

Brams, S. J. (1978). *The Presidential Election Game*. New Haven: Yale University Press.

Brams, S. J. (1980). *Biblical Games: A Strategic Analysis of Stories in the Old Testament*. Cambridge, Mass.: The MIT Press.

Brayer, A. R. (1964). An experimental analysis of some variables in minimax theory. *Behavioral Science*, **9**, 33–44.

Brockmann, H. J., Grafen, A., and Dawkins, R. (1979). Evolutionarily stable nesting strategy in a digger wasp. *Journal of Theoretical Biology*, **77**, 473–96.

Brockner, J., Rubin, J. Z., and Lang, E. (1981). Face-saving and entrapment. *Journal of Experimental Social Psychology*, **17**, 68–79.

Brockner, J., Shaw, M. C., and Rubin, J. Z. (1979). Factors affecting withdrawal from an escalating conflict: Quitting before it's too late. *Journal of Experimental Social Psychology*, **15**, 492–503.

Burhans, D. T., Jr. (1973). Coalition game research: a reexamination. *American Journal of Sociology*, **79**, 389–408.

Caldwell, M. D. (1976). Communication and sex effects in a five-person Prisoner's Dilemma Game. *Journal of Personality and Social Psychology*, **33**, 273–80.

Carey, M. S. (1980). Real life and the dollar auction. In A. I. Teger, *Too Much Invested to Quit*, pp. 117–33. New York: Pergamon.

Chaney, M. V. and Vinacke, W. E. (1960). Achievement and nurturance in triads varying in power distribution. *Journal of Abnormal and Social Psychology*, **60**, 175–81.

Chernoff, H. (1954). Rational selection of decision functions. *Econometrica*, **22**, 423–43.

Chertkoff, J. M. (1966). The effects of probability of success on coalition formation. *Journal of Experimental Social Psychology*, **2**, 265–77.

Chertkoff, J. M. (1970). Social psychological theories and research in coalition formation. In S. Groennings, E. W. Kelley, and M. Lieserson (Eds), *The Study of Coalition Behaviors*, pp. 297–322. New York: Holt.

Chertkoff, J. M. (1971). Coalition formation as a function of differences in resources. *Journal of Conflict Resolution*, **15**, 371–83.

Chertkoff, J. M. and Braden, J. L. (1974). Effects of experience and bargaining restrictions on coalition formation. *Journal of Personality and Social Psychology*, **30**, 169–177.

Chertkoff, J. M. and Esser, J. K. (1976). A review of experiments in explicit bargaining. *Journal of Experimental Social Psychology*, **12**, 464–86.

Clutton-Brock, T. H. and Harvey, P. H. (Eds) (1978). *Readings in Sociobiology*. San Francisco: W. H. Freeman.

Cohen, J. M. and Cohen, M. J. (1980). *The Penguin Dictionary of Modern Quotations*. Harmondsworth: Penguin.

Collard, D. (1978). *Altruism and Economy: A Study in Non-Selfish Economics*. Oxford: Martin Robertson.

Colman, A. M. (1975). The psychology of influence. In B. Frost (Ed.), *The Tactics of Pressure*, pp. 11–24. London: Stainer & Bell.

Colman, A. M. (1979a). Abstract and Lifelike Experimental Games. Unpublished doctoral dissertation, Rhodes University.

Colman, A. M. (1979b). The Borda effect: evidence from small groups and from a British General Election. *Bulletin of the British Psychological Society*, **32**, 221 (Abstract).

Colman, A. M. (1980). The likelihood of the Borda effect in small decision-making committees. *British Journal of Mathematical and Statistical Psychology*, **33**, 50–6.

Colman, A. M. (1981). *What is Psychology?* London: Kogan Page.

Colman, A. M. (Ed.). (1982). *Cooperation and Competition in Humans and Animals*. Wokingham: Van Nostrand.

Colman, A. M. and Pountney, I. (1975a). Voting paradoxes: a Socratic dialogue. I. *Political Quarterly*, **46** (2), 186–90.

Colman, A. M. and Pountney, I. (1975b). Voting paradoxes: a Socratic dialogue. II. *Political Quarterly*, **46** (3), 304–9.

Colman, A. M. and Pountney, I. (1978). Borda's voting paradox: theoretical likelihood and electoral occurrences. *Behavioral Science*, **23**, 15–20.

Condorcet, M. J. A. N. C. Marquis de (1785). *Essai sur l'application de l'analyse à la probabilité des décisions rendues à la pluralité des voix*. In M. J. A. N. C. Marquis de Condorcet, *Oeuvres Complètes*. Paris, 1804.

Coombs, C. H. (1964). *A Theory of Data*. New York: Wiley.

Copeland, A. H. (1945). Review of von Neumann and Morgenstern's *Theory of Games and Economic Behavior*. *Bulletin of the American Mathematical Society*, **51**, 498–504.

Crosbie, P. V. and Kullberg, V. K. (1973). Minimum resource or balance in coalition formation. *Sociometry*, **36**, 476–93.

Crumbaugh, C. M. and Evans, G. W. (1967). Presentation format, other-person strategies, and cooperative behavior in the Prisoner's Dilemma. *Psychological Reports*, **20**, 895–902.

Davenport, W. (1960). Jamaican fishing: a game theory analysis. In S. W. Mintz (Ed.), *Papers in Caribbean Anthropology Nos. 57–64*, pp. 3–11. New Haven, Conn.: Yale University Publications in Anthropology.

David, F. N. (1962). *Games, Gods and Gambling: A History of Probability and Statistical Ideas*. London: Griffin.

Davis, J. H., Laughlin, P. R., and Komorita, S. S. (1976). The social psychology of small groups. *Annual Review of Psychology*, **27**, 501–542.

Davis, M. D. (1970). *Game Theory: A Nontechnical Introduction*. New York: Basic Books.

Dawes, R. M. (1973). The commons dilemma game: An *n*-person mixed-motive game with a dominating strategy for defection. *Oregon Research Institute Research Bulletin*, **13** (2).

Dawes, R. M. (1975). Formal models of dilemmas in social decision making. In M. Kaplan and S. Schwartz (Eds), *Human Judgement and Decision Processes: Formal and Mathematical Approaches*, pp. 87–107. New York: Academic Press.

Dawes, R. M. (1980). Social dilemmas. *Annual Review of Psychology*, **31**, 169–93.

Dawes, R. M., McTavish, J., and Shaklee, H. (1977). Behavior, communication, and assumptions about other people's behavior in a Commons Dilemma situation. *Journal of Personality and Social Psychology*, **35**, 1–11.

Dawkins, R. (1976). *The Selfish Gene*. Oxford: Oxford University Press.

Dawkins, R. and Brockmann, H. J. (1980). Do digger wasps commit the Concorde fallacy? *Animal Behaviour*, **28**, 892–6.

Dawkins, R. and Carlisle, T. R. (1976). Parental investment, mate desertion, and a fallacy. *Nature*, **262**, 131–3.

DeMeyer, F. and Plott, C. (1970). The probability of a cylical majority. *Econometrica*, **38**, 345–54.

Deutsch, M. (1958). Trust and suspicion. *Journal of Conflict Resolution*, **2**, 265–79.

Deutsch, M. (1969). Socially relevant science: reflections on some studies of interpersonal conflict. *American Psychologist*, **24**, 1076–92.

Deutsch, M. and Krauss, R. M. (1960). The effect of threat upon interpersonal bargaining. *Journal of Abnormal and Social Psychology*, **61**, 181–9.

Deutsch, M. and Krauss, R. M. (1962). Studies of interpersonal bargaining. *Journal of Conflict Resolution*, **6**, 52–76.

Dodgson, C. L. (1876). *A Method of Taking Votes on More than Two Issues*. Oxford: Clarendon. Reprinted in Black (1958).

Downs, A. (1957). *An Economic Theory of Democracy*. New York: Harper & Row.

Druckman, D. (Ed.). (1977). *Negotiations: Social Psychological Perspectives*. New York: Sage-Halsted.

Edgeworth, F. Y. (1881). *Mathematical Psychics*. London: Kegan Paul.

Edwards, W. and Tversky, A. (Eds) (1967). *Decision Making: Selected Readings*. Harmondsworth: Penguin.

Eibl-Eibesfeldt, I. (1970). *Ethology: The Biology of Behavior*. New York: Holt, Rinehart, & Winston.

Einhorn, H. J. and Hogarth, R. M. (1981). Behavioral decision theory: processes of judgment and choice. *Annual Review of Psychology*, **32**, 53–88.

Eiser, J. R. (1980). *Cognitive Social Psychology: A Guidebook to Theory and Research.* Maidenhead: McGraw-Hill.

Eiser, J. R. and Bhavnani, K.-K. (1974). The effect of situational meaning on the behaviour of subjects in the Prisoner's Dilemma Game. *European Journal of Social Psychology*, **4**, 93–7.

Eiser, J. R. and Tajfel, H. (1972). Acquisition of information in dyadic interaction. *Journal of Personality and Social Psychology*, **23**, 340–5.

Evans, G. (1964). Effect of unilateral promise and value of rewards upon cooperation and trust. *Journal of Abnormal and Social Psychology*, **69**, 587–90.

Evans, G. and Crumbaugh, C. M. (1966). Effects of prisoner's dilemma format on cooperative behavior. *Journal of Personality and Social Psychology*, **3**, 486–8.

Farquharson, R. (1968). *Drop Out!* Harmondsworth: Penguin.

Farquharson, R. (1969). *Theory of Voting.* Oxford: Basil Blackwell.

Festinger, L. (1957). *A Theory of Cognitive Dissonance.* New York: Row, Peterson.

Fishburn, P. C. (1970). Arrow's impossibility theorem: concise proof and finite voters. *Journal of Economic Theory*, **2**, 103–6.

Fishburn, P. C. (1973a). *The Theory of Social Choice.* Princeton, N.J.: Princeton University Press.

Fishburn, P. C. (1973b). Voter concordance, simple majorities, and group decision methods. *Behavioral Science*, **18**, 364–76.

Fishburn, P. C. (1974a). Simple voting systems and majority rule. *Behavioral Science*, **19**, 166–76.

Fishburn, P. C. (1974b). Single-peaked preferences and probabilities of cyclical majorities. *Behavioral Science*, **19**, 21–7.

Fishburn, P. C. (1977). An analysis of voting procedures with nonranked voting. *Behavioral Science*, **22**, 178–85.

Fishburn, P. C. and Gehrlein, W. V. (1976). An analysis of simple two-stage voting systems. *Behavioral Science*, **21**, 1–12.

Fisher, R. A. (1930). *The Genetical Theory of Natural Selection.* Oxford: Oxford University Press.

Fisher, R. A. (1934). Randomisation, and an old enigma of card play. *Mathematical Gazette*, **18**, 294–7.

Fletcher, J. (1966). *Situation Ethics: The New Morality.* Philadelphia: Westminster Press.

Flood, M. M. (1958). Some experimental games. *Management Science*, **5**, 5–26.

Fox, J. (1972). The learning of strategies in a simple, two-person zero-sum game without saddlepoint. *Behavioral Science*, **17**, 300–8.

Fox, J. and Guyer, M. (1973). Equivalence and stooge strategies in zero-sum games. *Journal of Conflict Resolution*, **17**, 513–33.

Fox, J. and Guyer, M. (1977). Group size and others' strategy in an N-person game. *Journal of Conflict Resolution*, **21**, 323–38.

Fox, J. and Guyer, M. (1978). "Public" choice and cooperation in N-Person Prisoner's Dilemma. *Journal of Conflict Resolution*, **22**, 469–81.

Franklin, J. (1980). *Methods of Mathematical Economics: Linear and Nonlinear Programming, Fixed-Point Theorems.* New York: Springer-Verlag.

Fréchet, M. (1953). Émile Borel, initiator of the theory of psychological games and its application. *Econometrica*, **21**, 95–6.

Gallo, P. S. and McClintock, C. G. (1965). Cooperative and competitive behavior in mixed-motive games. *Journal of Conflict Resolution*, **9**, 68–78.

Gallo, P. S. (1966). Effects of increased incentives upon the use of threat in bargaining. *Journal of Personality and Social Psychology*, **4**, 14–20.

Gamson, W. A. (1961a). An experimental test of a theory of coalition formation. *American Sociological Review*, **26**, 565–73.

Gamson, W. A. (1961b). A theory of coalition formation. *American Sociological Review*, **26**, 373–82.

Gamson, W. A. (1962). Coalition formation at presidential nominating conventions. *American Journal of Sociology*, **68**, 157–71.

Gamson, W. A. (1964). Experimental studies of coalition formation. In L. Berkowitz (Ed.), *Advances in Experimental Social Psychology*. Vol. 1, pp. 82–110. New York: Academic Press.

Gärdenfors, P. (1976). Manipulation of social choice functions. *Journal of Economic Theory*, **13**, 217–28.

Gardner, M. (1974). Mathematical games. *Scientific American*, **231** (4), 120–4.

Garman, M. and Kamien, M. (1968). The paradox of voting: Probability calculations. *Behavioral Science*, **13**, 306–16.

Geist, V. (1966). The evolution of horn-like organs. *Behaviour*, **27**, 175–214.

Geist, V. (1974). On fighting strategies in animal combat. *Nature*, **250**, 354.

Geist, V. (1978). On weapons, combat, and ecology. In L. Krames, P. Pliner, and T. Alloway (Eds), *Aggression, Dominance, and Individual Spacing*. New York: Plenum.

Gibbard, A. (1973). Manipulation of voting schemes: a general result. *Econometrica*, **41**, 587–601.

Gibbs, J. (1982). Sex and Communication Effects in a Mixed-Motive Game. Unpublished doctoral dissertation, University of Durham.

Gleser, L. J. (1969). The paradox of voting: some probabilistic results. *Public Choice*, **7**, 47–64.

Goehring, D. J. and Kahan, J. P. (1976). The uniform N-Person Prisoner's Dilemma Game: construction and test of an index of cooperation. *Journal of Conflict Resolution*, **20**, 111–28.

Goodnow, J. J. (1955). Determinants of choice-distribution in two-choice situations. *American Journal of Psychology*, **68**, 106–16.

Grodzins, M. (1957). Metropolitan segregation. *Scientific American*, **197** (4), 33–41.

Grofman, B. (1972). A note on some generalizations of the paradox of cyclical majorities. *Public Choice*, **12**, 113–14.

Guilbaud, G. T. (1952). Les théories de l'interêt général et la problème logique de l'aggregation. *Economie Appliquée*, **5**, 501–584. Trans. and repr. in P. F. Lazarsfeld and N. W. Henry (Eds), *Readings in Mathematical Social Sciences*, pp. 262–307. Chicago: Science Research Associates.

Gumpert, P., Deutsch, M., and Epstein, Y. (1969). Effects of incentive magnitude on cooperation in the Prisoner's Dilemma game. *Journal of Personality and Social Psychology*, **11**, 66–9.

Guyer, M. and Perkel, B. (1972). Experimental games: a bibliography (1945–1971). Ann Arbor, Mich.: Mental Health Research Institute, Communication 293.

Guyer, M. and Rapoport, A. (1972). 2 × 2 games played once. *Journal of Conflict Resolution*, **16**, 409–31.

Hamburger, H. (1973). N-person Prisoner's Dilemma. *Journal of Mathematical Sociology*, **3**, 27–48.

Hamburger, H. (1977). Dynamics of cooperation in take-some games. In W. H. Kempf and B. H. Repp (Eds), *Mathematical Models for Social Psychology*. Bern: Hans Huber.

Hamburger, H. (1979). *Games as Models of Social Phenomena*. San Francisco: W. H. Freeman.

Hamburger, H., Guyer, M., and Fox, J. (1975). Group size and cooperation. *Journal of Conflict Resolution*, **19**, 503–31.

Hamilton, W. D. (1967). Extraordinary sex ratios. *Science*, **156**, 477–88.

Hamner, W. C. (1977). The influence of structural, individual and strategic differences on bargaining outcomes: a review. In D. L. Harnett and L. L. Cummings (Eds), *Bargaining Behavior and Personality*. Bloomington: University of Indiana Press.

Hansson, B. (1976). The existence of group preferences. *Public Choice*, **28**, 89–98.

Hardin, G. (1968). The tragedy of the commons. *Science*, **162**, 1243–8.

Harford, T. C. and Solomon, L. (1967). "Reformed sinner" and "lapsed saint" strategies in the Prisoner's Dilemma game. *Journal of Conflict Resolution*, **11**, 104–9.

Harris, R. J. (1968). Note on Howard's "Theory of meta-games". *Psychological Reports*, **24**, 849–50.

Harris, R. J. (1969). A geometric classification system for 2 × 2 interval-symmetric games. *Behavioral Science*, **14**, 138–46.

Harris, R. J. (1972). An interval-scale clasification system for all 2 × 2 games. *Behavioral Science*, **17**, 371–83.

Harvey, J. H. and Smith, W. P. (1977). *Social Psychology: An Attributional Approach*. St Louis: Mosby.

Haywood, O. G., Jr. (1954). Military decision and game theory. *Journal of the Operations Research Society of America*, **2**, 365–85.

Hillinger, C. (1971). Voting on issues and on platforms. *Behavioral Science*, **16**, 564–6.

Hollis, M. (1979). Rational man and social science. In R. Harrison (Ed.), *Rational Action: Studies in Philosophy and Social Science*. Cambridge: Cambridge University Press.

Hottes, J. and Kahn, A. (1974). Sex differences in a mixed-motive conflict situation. *Journal of Personality*, **42**, 260–75.

Howard, N. (1971). *Paradoxes of Rationality: Theory of Metagames and Political Behavior*. Cambridge, Mass.: MIT Press.

Howard, N. (1974). "General" metagames: An extension of the metagame concept. In A. Rapoport (Ed.), *Game Theory as a Theory of Conflict Resolution*, pp. 261–83. Dordrecht: D. Reidel.

Humphreys, L. G. (1939). Acquisition and extinction of verbal expectations in a situation analogous to conditioning. *Journal of Experimental Psychology*, **25**, 294–301.

Jerdee, T. H. and Rosen, B. (1974). Effects of opportunity to communicate and visibility of individual decisions on behavior in the common interest. *Journal of Applied Psychology*, **59**, 712–16.

Jones, A. J. (1980). *Game Theory: Mathematical Models of Conflict*. Chichester: Ellis Horwood.

Joyce, R. C. (1976). Sophisticated Voting for Three Candidate Contests Under the Plurality Rule. Unpublished doctoral dissertation, University of Rochester.

Kahan, J. P. (1973). Noninteraction in an anonymous three-person Prisoner's Dilemma Game. *Behavioral Science*, **18**, 124–7.

Kahan, J. P. and Goehring, D. J. (1973). Responsiveness in two-person zero-sum games. *Behavioral Science*, **18**, 27–33.

Kahan, J. P. and Rapoport, A. (1974). Decisions of timing in bipolarized conflict situations with complete information. *Acta Psychologica*, **38**, 183–203.

Kahn, H. (1965). *On Escalation*. New York: Praeger.

Kahneman, D. and Tversky, A. (1982). The psychology of preferences. *Scientific American*, **246** (1), 136–42.

Kalisch, G. K., Milnor, J. W., Nash, J. F., and Nering, E. D. (1954). Some experimental *n*-person games. In R. M. Thrall, C. H. Coombs, and R. L. Davis (Eds), *Decision Processes*, pp. 301–27. New York: Wiley.

Kanouse, D. E. and Wiest, W. M. (1967). Some factors affecting choice in the Prisoner's Dilemma. *Journal of Conflict Resolution*, **11**, 206–13.

Kant, I. (1785). *Grundlegung zur Metaphysik der Sitten*. In E. Cassirer (Ed.), (1912–22) *Immanuel Kants Werke*, Vol. IV. Berlin.

Karlin, S. (1959). *Mathematical Models and Theory in Games, Programming, and Economics*. Vol. 2. Reading: Addison-Wesley.

Kaufman, H. and Becker, G. M. (1961). The empirical determination of game-theoretical strategies. *Journal of Experimental Psychology*, **61**, 462–8.

Kaufman, H. and Lamb, J. C. (1967). An empirical test of game theory as a descriptive model. *Perceptual and Motor Skills*, **24**, 951–60.

Kelley, H. H. (1965). Experimental studies of threats in interpersonal negotiations. *Journal of Conflict Resolution*, **9**, 79–105.

Kelley, H. H. and Arrowood, A. J. (1960). Coalitions in the triad: Critique and experiment. *Sociometry*, **23**, 231–44.

Kelley, H. H. and Grzelak, J. L. (1972). Conflict between individual and common interest in an *n*-person relationship. *Journal of Personality and Social Psychology*, **21**, 190–7.

Kelley, H. H. and Stahelski, A. J. (1970a). Errors in perception of intentions in a mixed-motive game. *Journal of Experimental Social Psychology*, **6**, 379–400.

Kelley, H. H. and Stahelski, A. J. (1970b). The inference of intentions from moves in the Prisoner's Dilemma Game. *Journal of Experimental Social Psychology*, **6**, 401–19.

Kelley, H. H. and Stahelski, A. J. (1970c). Social interaction basis of cooperators' and competitors' beliefs about others. *Journal of Personality and Social Psychology*, **16**, 66–91.

Kelley, H. H., Thibaut, J. W., Radloff, R., and Mundy, D. (1962). The development of cooperation in the "minimal social situation". *Psychological Monographs*, **76**, Whole No. 19.

Kelly, J. S. (1974). Voting anomalies, the number of voters, and the number of alternatives. *Econometrica*, **42**, 239–51.

Kelly, J. S. (1978). *Arrow Impossibility Theorems*. New York: Academic Press.

Keynes, J. M. (1936). *The General Theory of Employment, Interest and Money*. London: Macmillan.

Kirman, A. P. and Sondermann, D. (1972). Arrow's theorem, many agents, and invisible dictators. *Journal of Economic Theory*, **5**, 267–77.

Komorita, S. S. and Chertkoff, J. M. (1973). A bargaining theory of coalition formation. *Psychological Review*, **80**, 149–62.

Kozelka, R. (1969). A Bayesian approach to Jamaican fishing. In I. R. Buchler and H. G. Nutini (Eds), *Game Theory in the Behavioral Sciences*, pp. 117–25. Pittsburgh: University of Pittsburgh Press.

Kramer, G. H. (1972). Sophisticated voting over multidimensional choice spaces. *Journal of Mathematical Sociology*, **2**, 165–80.

Kramer, G. H. (1973). On a class of equilibrium conditions for majority rule. *Econometrica*, **41**, 285–97.

Kugn, K. and Nagatani, K. (1974). Voter antagonism and the paradox of voting. *Econometrica*, **42**, 1045–67.

Kuhlman, D. M. and Wimberley, D. L. (1976). Expectations of choice behavior held by cooperators, competitors, and individualists across four classes of experimental game. *Journal of Personality and Social Psychology*, **34**, 69–81.

Lacey, O. L. and Pate, J. C. (1960). An experimental test of game theory as a descriptive model. *Psychological Reports*, **7**, 527–30.

Lamm, H. and Myers, D. G. (1978). Group-induced polarization of attitudes and behavior. *Advances in Experimental Social Psychology*, **11**, 145–95.

Laplace, P.-S. (1814). Théorie Analytique des Probabilités. 2nd ed. Paris. In P.-S. Laplace, (1878–1912) *Oeuvres Complètes*, Vol. 7, Supplement 1. Paris.

Lazarus, J. (1982). Competition and conflict in animals. In A. M. Colman (Ed.), *Cooperation and Competition in Humans and Animals*, pp. 26–56. Wokingham: Van Nostrand.

Leavitt, H. J. (1951). Some effects of certain communication patterns on group performance. *Journal of Abnormal and Social Psychology*, **46**, 38–50.

Lewontin, R. C. (1961). Evolution and the theory of games. *Journal of Theoretical Biology*, **1**, 382–403.

Lieberman, B. (1959). Human behavior in a strictly determined 2 × 2 matrix game. *Research Memorandum*, Department of Social Relations, Harvard University.

Lieberman, B. (1960a). A failure of game theory to predict human behavior. *Memo SP-101*, Laboratory of Social Relations, Harvard University.

Lieberman, B. (1960b). Human behavior in a strictly determined 3 × 3 matrix game. *Behavioral Science*, **5**, 317–22.

Lieberman, B. (1962). Experimental studies of conflict in some two-person and three-person games. In J. H. Criswell, H. Solomon, and P. Suppes (Eds), *Mathematical Models in Small Group Processes*, pp. 203–20. Stanford: Stanford University Press.

Lloyd, W. F. (1833). *Two Lectures on the Checks to Population*. Oxford: Oxford University Press.

Lorenz, K. (1966). *On Aggression*. London: Methuen.

Lucas, W. F. (1968). The proof that a game may not have a solution. *Memorandum RM-5543-PR*. Santa Monica, Ca.: The Rand Corporation.

Luce, R. D. and Raiffa, H. (1957). *Games and Decisions: Introduction and Critical Survey*. New York: Wiley.

Lumsden, M. (1973). The Cyprus conflict as a Prisoner's Dilemma Game. *Journal of Conflict Resolution*, **17**, 7–32.

Mack, D. (1975). Skirting the competition. *Psychology Today*, **8**, 39–41.

Mackie, J. L. (1978). The law of the jungle: Moral alternatives and principles of evolution. *Philosophy*, **53**, 455–64.

Mackie, J. L. (1982). Cooperation, competition, and moral philosophy. In A. M. Colman (Ed.), *Cooperation and Competition in Humans and Animals*, pp. 271–84. Wokingham: Van Nostrand.

Magenau, J. M. and Pruitt, D. G. (1979). The social psychology of bargaining. In G. M. Stephenson and C. J. Brotherton (Eds), *Industrial Relations: A Social Psychological Approach*. Chichester: Wiley.

Malcolm, D. and Lieberman, B. (1965). The behavior of responsive individuals playing a two-person, zero-sum game requiring the use of mixed strategies. *Psychonomic Science*, **2**, 373–4.

Mann, I. and Shapley, L. S. (1964). The a priori voting strength of the electoral college. In M. Shubik (Ed.), *Game Theory and Related Approaches to Social Behavior*, pp. 151–64. New York: Wiley.

Martin, D. (1978). The selective usefulness of game theory. *Social Studies of Science SAGE*, **8**, 85–110.

Marwell, G. and Ames, R. E. (1979). Experiments on the provision of public goods I: Resources, interest, group size, and the free rider problem. *American Journal of Sociology*, **84**, 1335–60.

Marwell, G. and Schmitt, D. (1972). Cooperation in a three-person Prisoner's Dilemma. *Journal of Personality and Social Psychology*, **21**, 376–83.

May, R. M. (1971). Some mathematical remarks on the paradox of voting. *Behavioral Science*, **16**, 143–51.

Maynard Smith, J. (1972). Game theory and the evolution of fighting. In J. Maynard Smith, *On Evolution*, pp. 8–28. Edinburgh: Edinburgh University Press.

Maynard Smith, J. (1974). The theory of games and the evolution of animal conflicts. *Journal of Theoretical Biology*, **47**, 209–21.

Maynard Smith, J. (1976a). Evolution and the theory of games. *American Scientist*, **64** (1), 41–5.

Maynard Smith, J. (1976b). Group selection. *Quarterly Review of Biology*, **51**, 277–83.

Maynard Smith, J. (1978a). The evolution of behavior. *Scientific American*, **239** (3), 136–45.

Maynard Smith, J. (1978b). Optimization theory in evolution. *Annual Review of Ecology and Systematics*, **9**, 31–56.

Maynard Smith, J. (1979). Game theory and the evolution of behaviour. *Proceedings of the Royal Society of London B*, **205**, 475–88.

Maynard Smith, J. and Price, G. R. (1973). The logic of animal conflict. *Nature*, **246**, 15–18.

McClintock, C. G. (1972). Game behavior and social motivation in interpersonal settings. In C. G. McClintock (Ed.), *Experimental Social Psychology*, pp. 271–97. New York: Holt, Rinehart, & Winston.

McClintock, C. G. and McNeel, S. P. (1967). Prior dyadic experience and monetary reward as determinants of cooperative behavior. *Journal of Personality and Social Psychology*, **5**, 282–94.

McKelvey, R. D. and Niemi, R. G. (1978). A multistage game representation of sophisticated voting for binary procedures. *Journal of Economic Theory*, **18**, 1–22.

McVicar, J. (1981). Postscript. In S. Cohen and L. Taylor, *Psychological Survival: The Experience of Long-Term Imprisonment*, pp. 221–9. 2nd ed. Harmondsworth: Penguin.

Mertens, J.-F. and Zamir, S. (1980). Minimax and maximin of repeated games with incomplete information. *International Journal of Game Theory*, **9**, 201–15.

Messick, D. M. (1967). Interdependent decision strategies in zero-sum games: a computer-controlled study *Behavioral Science*, **12**, 33–48.

Messick, D. M. (1973). To join or not to join: an approach to the unionization decision. *Organizational Behavior and Human Performance*, **10**, 145–56.

Michener, H. A., Fleishman, J. A., Vaske, J. J., and Statza, G. R. (1975). Minimum resource and pivotal power theories: A competitive test in 4-person coalition situations. *Journal of Conflict Resolution*, **19**, 89–107.

Miller, D. R. (1973). A Shapley value analysis of the proposed Canadian constitutional amendment scheme. *Canadian Journal of Political Science*, **6**, 140–3.

Miller, D. T. and Holmes, J. G. (1975). The role of situational restrictiveness on self-fulfilling prophecies: a theoretical and empirical extension of Kelley and Stahelski's triangle hypothesis. *Journal of Personality and Social Psychology*, **31**, 661–73.

Miller, N. R. (1977). Graph-theoretic approaches to the theory of voting. *American Journal of Political Science*, **21**, 769–803.

Milnor, J. (1954). Games against nature. In R. M. Thrall, C. H. Coombs, and R. L. Davis (Eds.), *Decision Processes*, pp. 49–59. New York: Wiley.

Mintz, A. (1951). Non-adaptive group behavior. *Journal of Abnormal and Social Psychology*, **46**, 150–9.

Moore, P. G. (1980). *Reason by Numbers*. Harmondsworth: Penguin.

Morin, R. E. (1960). Strategies in games with saddle-points. *Psychological Reports*, **7**, 479–83.

Morley, I. E. (1981). Negotiation and bargaining. In M. Argyle (Ed.), *Social Skills and Work*, pp. 84–115. London: Methuen.

Morley, I. E. and Stephenson, G. M. (1977). *The Social Psychology of Bargaining*. London: George Allen & Unwin.

Myers, D. G. and Lamm, H. (1976). The group polarization phenomenon. *Psychological Bulletin*, **83**, 602–27.

Nanson, E. J. (1882). Methods of elections. *Transactions and Proceedings of the Royal Society of Victoria*, **18**.

Nash, J. F. (1950). Equilibrium points in *n*-person games. *Proceedings of the National Academy of Sciences, U.S.A.*, **36**, 48–9.

Nash, J. F. (1951). Non-cooperative games. *Annals of Mathematics*, **54**, 286–95.

Nemeth, C. (1970). Bargaining and reciprocity. *Psychological Bulletin*, **74**, 297–308.

Nemeth, C. (1972). A critical analysis of research utilizing the prisoner's dilemma paradigm for the study of bargaining. *Advances in Experimental Social Psychology*, **6**, 203–34.

Nemeth, C. (Ed.) (1974). *Social Psychology: Classic and Contemporary Integrations*. Chicago: Rand McNally.

Niemi, R. G. (1969). Majority decision-making with partial unidimensionality. *American Political Science Review*, **63**, 488–97.

Niemi, R. G. (1970). The occurrence of the paradox of voting in university elections. *Public Choice*, **8**, 91–100.

Niemi, R. G. (1982). An exegesis of Farquharson's *Theory of Voting*. *Public Choice* (In press).

Niemi, R. G. and Frank, A. Q. (1980). Sophisticated voting under the plurality procedure. Paper prepared for a conference in honour of W. H. Riker, Washington D.C., 27 August.

Niemi, R. G. and Riker, W. H. (1976). The choice of voting systems. *Scientific American*, **234** (6), 21–7.

Niemi, R. G. and Weisberg, H. F. (1968). A mathematical solution for the probability of the paradox of voting. *Behavioral Science*, **13**, 317–23.

Niemi, R. G. and Weisberg, H. F. (Eds) (1972). *Probability Models of Collective Decision Making*. Columbus: Charles E. Merrill.

Niemi, R. G. and Weisberg, H. F. (1973). A pairwise probability approach to the likelihood of the paradox of voting. *Behavioral Science*, **18**, 109–18.

Ofshe, L. and Ofshe, R. (1970). *Utility and Choice in Social Interaction*. Englewood Cliffs, N.J.: Prentice-Hall.

Orwant, C. J. and Orwant, J. E. (1970). A comparison of interpreted and abstract versions of mixed-motive games. *Journal of Conflict Resolution*, **14**, 91–7.

Osgood, C. E. (1962). *An Alternative to War or Surrender*. Urbana, Ill.: University of Illinois Press.

Oskamp, S. (1971). Effects of programmed strategies on cooperation in the Prisoner's Dilemma and other mixed-motive games. *Journal of Conflict Resolution*, **15**, 225–59.

Oskamp, S. and Kleinke, C. (1970). Amount of reward as a variable in the Prisoner's Dilemma game. *Journal of Personality and Social Psychology*, **16**, 133–40.

Owen, G. (1968). *Game Theory*. Philadelphia: W. B. Saunders.

Parfit, D. (1979a). Is common-sense morality self-defeating? *Journal of Philosophy*, **76**, 533–45.

Parfit, D. (1979b). Prudence, morality, and the Prisoner's Dilemma. *Proceedings of the British Academy*, **65**, 539–64.

Parker, G. A. (1970). The reproductive behaviour and nature of sexual selection in *Scatophaga stercoraria* L. II. *Journal of Animal Ecology*, **39**, 205–28.

Parker, G. A. (1978). Selfish genes, evolutionary games, and the adaptiveness of behaviour. *Nature*, **274**, 849–55.

Pate, J. L. (1967). Losing games and their recognition. *Psychological Reports*, **20**, 1031–5.

Pate, J. L. and Broughton, E. D. (1970). Game-playing behavior as a function of incentive. *Psychological Reports*, **27**, 36.

Pate, J. L., Broughton, E. D., Hallman, L. K., and Letterman, A. N. L. (1974). Learning in two-person zero-sum games. *Psychological Reports*, **34**, 503–10.

Pattanaik, P. K. (1976). Counter-threats and strategic manipulation under voting schemes. *Review of Economic Studies*, **43**, 11–18.

Payne, W. (1965). Acquisition of strategies in gaming situations. *Perceptual and Motor Skills*, **20**, 473–9.

Pinter, H. (1960). *The Caretaker*. London: Methuen.

Pinter, H. (1972). Harold Pinter: An interview. In A. Ganz (Ed.), *Pinter: A Collection of Critical Essays*, pp. 19–33. Englewood Cliffs, N.J.: Prentice-Hall.

Plon, M. (1974). On the meaning of the notion of conflict and its study in social psychology. *European Journal of Social Psychology*, **4**, 389–436.

Popper, K. R. (1961). Evolution and the Tree of Knowledge. Herbert Spencer Lecture delivered in Oxford, 30 October. Reprinted in K. R. Popper, *Objective Knowledge: An Evolutionary Approach*, pp. 256–84. Oxford: Clarendon, 1972.

Popper, K. R. (1969). *Conjectures and Refutations: The Growth of Scientific Knowledge*. 3rd ed. London: Routledge & Kegan Paul.

Pruitt, D. G. (1967). Reward structure and cooperation: The decomposed prisoner's dilemma game. *Journal of Personality and Social Psychology*, **7**, 21–7.

Pruitt, D. G. (1970). Motivational processes in the decomposed prisoner's dilemma game. *Journal of Personality and Social Psychology*, **14**, 227–38.

Pruitt, D. G. (1976). Power and bargaining. In B. Seidenberg and A. Snadowsky (Eds), *Social Psychology: An Introduction*, pp. 343–75. New York: Free Press.

Pruitt, D. G. and Kimmel, M. J. (1977). Twenty years of experimental gaming: Critique, synthesis, and suggestions for the future. *Annual Review of Psychology*, **28**, 363–92.

Psathas, G. and Stryker, S. (1965). Bargaining behavior and orientations in coalition formation. *Sociometry*, **28**, 124–44.

Rabinowitz, K., Kelley, H. H., and Rosenblatt, R. M. (1966). Effects of different types of interdependence and response conditions in the minimal social situation. *Journal of Experimental Social Psychology*, **2**, 169–97.

Rapoport, A. (1962). The use and misuse of game theory. *Scientific American*, **207** (6), 108–18.

Rapoport, A. (1967a). Escape from paradox. *Scientific American*, **217** (1), 50–6.

Rapoport, A. (1967b). Exploiter, Leader, Hero, and Martyr: the four archetypes of the 2 × 2 game. *Behavioral Science*, **12**, 81–4.

Rapoport, A. (1970). Conflict resolution in the light of game theory and beyond. In P. Swingle (Ed.), *The Structure of Conflict*, pp. 1–42. New York: Academic Press.

Rapoport, A. (1971). *The Big Two: Soviet-American Perceptions of Foreign Policy*. Indianapolis: Pegasus.

Rapoport, A. and Chammah, A. M. (1965). *Prisoner's Dilemma: A Study in Conflict and Cooperation*. Ann Arbor: University of Michigan Press.

Rapoport, A. and Chammah, A. M. (1969). The game of Chicken. In I. R. Buchler and H. G. Nutini (Eds), *Game Theory in the Behavioral Sciences*, pp. 151–75. Pittsburgh: University of Pittsburgh Press.

Rapoport, A. and Guyer, M. (1966). A taxonomy of 2 × 2 games. *General Systems*, **11**, 203–14.

Rapoport, A. and Orwant, C. (1962). Experimental games: a review. *Behavioral Science*, **7**, 1–37.

Rapoport, A., Kahan, J. P., and Stein, W. E. (1976). Decisions of timing in experimental probabilistic duels. *Journal of Mathematical Psychology*, **13**, 163–91.

Rescher, N. (1975). *Unselfishness: The Role of the Vicarious Affects in Moral Philosophy and Social Theory*. Pittsburgh, Pa.: University of Pittsburgh Press.

Reychler, L. (1979). The effectiveness of a pacifist strategy in conflict resolution: an experimental study. *Journal of Conflict Resolution*, **23**, 228–60.

Rhinehart, L. (1972). *The Dice Man*. Frogmore: Panther.

Richelson, J. T. (1979). Soviet strategic doctrine and limited nuclear operations. *Journal of Conflict Resolution*, **23**, 326–36.

Riker, W. H. (1961). Voting and the summation of preferences: An interpretative bibliographical review of selected developments during the last decade. *American Political Science Review*, **55** (4), 900–11.

Riker, W. H. (1962). *The Theory of Political Coalitions*. New Haven, Conn.: Yale University Press.

Riker, W. H. (1965). Arrow's theorem and some examples of the paradox of voting. In J. M. Claunch (Ed.), *Mathematical Applications in Political Science*, pp. 41–60. Dallas: Southern Methodist University Press.

Riker, W. H. (1967). A new proof of the size principle. In J. L. Bernd (Ed.), *Mathematical Applications in Political Science II*, pp. 167–74. Dallas: Southern Methodist University Press.

Riker, W. H. and Ordeshook, P. C. (1973). *An Introduction to Positive Political Theory*. Englewood Cliffs, N.J.: Prentice-Hall.

Robinson, J. (1951). An iterative method of solving a game. *Annals of Mathematics*, **54**, 296–301.

Robinson, M. (1975). Prisoner's Dilemma: metagames and other solutions. *Behavioral Science*, **20**, 201–5.

Robinson, T. W. (1970). Game theory and politics: recent Soviet views. *Studies in Soviet Thought*, **10**, 291–315.

Rosenthal, H. (1974). Game-theoretic models of bloc-voting under proportional representation: really sophisticated voting in French labor elections. *Public Choice*, **18**, 1–23.

Rousseau, J.-J. (1762). *Du Contrat Social ou Principes du Droit Politique*. In J.-J. Rousseau, *Oeuvres Complètes*, Vol. III. Dijon: Éditions Gallimard, 1964.

Rubin, J. Z. and Brockner, J. (1975). Factors affecting entrapment in waiting situations: the Rosencrantz and Guildenstern effect. *Journal of Personality and Social Psychology*, **31**, 1054–63.

Rubin, J. Z., Brockner, J., Small-Weil, S., and Nathanson, S. (1980). Factors affecting entry into psychological traps. *Journal of Conflict Resolution*, **24**, 405–26.

Rubin, J. Z. and Brown, B. R. (1975). *The Social Psychology of Bargaining and Negotiation*. New York: Academic Press.

Runciman, W. G. and Sen, A. K. (1965). Games, justice and the general will. *Mind*, **74**, 554–62.

Russell, B. (1954). *Human Society in Ethics and Politics*. London: George Allen & Unwin.

Rustow, D. (1955). *The Politics of Compromise: A Study of Parties and Cabinet Government in Sweden*. Princeton, N.J.: Princeton University Press.

Sakaguchi, M. (1960). Reports on experimental games. *Statistical Applied Research JUSE*, **7**, 156–65.

Satterthwaite, M. A. (1973). The Existence of a Strategy Proof Voting Procedure: A Topic in Social Choice Theory. Unpublished doctoral dissertation, University of Wisconsin.

Satterthwaite, M. A. (1975). Strategy-proofness and Arrow's conditions: Existence and correspondence theorems for voting procedures and social welfare functions. *Journal of Economic Theory*, **10**, 187–217.

Savage, L. J. (1951). A theory of statistical decision. *Journal of the American Statistical Association*, **46**, 55–67.

Schelling, T. C. (1960). *The Strategy of Conflict*. Cambridge, Mass.: Harvard University Press.

Schelling, T. C. (1966). *Arms and Influence*. New Haven: Yale University Press.

Schelling, T. C. (1967). What is game theory? In J. C. Charlesworth (Ed.), *Contemporary Political Analysis*, pp. 224–32. New York: Free Press.

Schelling, T. C. (1968). Game theory and the study of ethical systems. *Journal of Conflict Resolution*, **12**, 34–44.

Schelling, T. C. (1971a). Dynamic models of segregation. *Journal of Mathematical Sociology*, **1**, 143–86.

Schelling, T. C. (1971b). On the ecology of micromotives. *The Public Interest*, **25**, 61–98.

Schelling, T. C. (1973). Hockey helmets, concealed weapons, and daylight saving: a study of binary choices with externalities. *Journal of Conflict Resolution*, **17**, 381–428.

Schlenker, B. R. and Bonoma, T. V. (1978). Fun and games: the validity of games for the study of conflict. *Journal of Conflict Resolution*, **22**, 7–38.

Schubert, G. A. (1959). *Quantitative Analysis of Judicial Behavior*. Glencoe, Ill.: Free Press.
Scientific American (1981). Program Power, **244** (4), 71A–72.
Scodel, A., Minas, J. S., Ratoosh, P., and Lipetz, M. (1959). Some descriptive aspects of two-person non-zero-sum games. *Journal of Conflict Resolution*, **3**, 114–19.
Sen, A. K. (1970). *Collective Choice and Social Welfare*. Edinburgh: Oliver & Boyd.
Séris, J.-P. (1974). *La Théorie des Jeux*. Paris: Presses Universitaires de France, Dossiers Logos.
Sermat, V. (1967a). The effect of initial cooperative or competitive treatment upon a subject's response to conditional cooperation. *Behavioral Science*, **12**, 301–13.
Sermat, V. (1967b). The possibility of influencing the other's behavior and cooperation: Chicken vs. Prisoner's Dilemma. *Canadian Journal of Psychology*, **27**, 204–19.
Sermat, V. (1970). Is game behavior related to behavior in other interpersonal situations? *Journal of Personality and Social Psychology*, **16**, 92–109.
Shapley, L. S. (1953). A value for *n*-person games. In H. W. Kuhn and A. W. Tucker (Eds), *Contributions to the Theory of Games, II*, pp. 307–17. Princeton, N.J.: Princeton University Press.
Shapley, L. S. and Shubik, M. (1954). A method of evaluating the distribution of power in a committee system. *American Political Science Review*, **48**, 787–92.
Shapley, L. S. and Snow, R. N. (1950). Basic solutions of discrete games. *Annals of Mathematical Studies*, **24**, 27–35.
Shepsle, K. A. (1972). Parties, voters, and the risk environment: A mathematical treatment of electoral competition under uncertainty. In R. G. Niemi and H. F. Weisberg (Eds), *Probability Models of Collective Decision Making*, pp. 273–97. Columbus, Ohio: Charles E. Merrill.
Sherif, M., Harvey, O. J., White, B. J., Hood, W. R., and Sherif, C. W. (1961). *Intergroup Conflict and Cooperation: The Robbers Cave Experiment*. Norman, Okla.: University of Oklahoma Book Exchange.
Shubik, M. (1954). Does the fittest necessarily survive? In M. Shubik (Ed.), *Readings in Game Theory and Political Behavior*, pp. 36–43. New York: Doubleday.
Shubik, M. (Ed.) (1964). *Game Theory and Related Approaches to Social Behavior*. New York: Wiley.
Shubik, M. (1971). The Dollar Auction game: A paradox in noncooperative behavior and escalation. *Journal of Conflict Resolution*, **15**, 109–11.
Shure, G. H., Meeker, R. J., and Hansford, E. A. (1965). The effectiveness of pacifist strategies in bargaining games. *Journal of Conflict Resolution*, **9**, 106–17.
Sidowski, J. B. (1957). Reward and punishment in a minimal social situation. *Journal of Experimental Psychology*, **54**, 318–26.
Sidowski, J. B., Wyckoff, L. B., and Tabory, L. (1956). The influence of reinforcement and punishment in a minimal social situation. *Journal of Abnormal and Social Psychology*, **52**, 115–19.
Siegel, S. and Goldstein, D. A. (1959). Decision-making behavior in a two-choice uncertain outcome situation. *Journal of Experimental Psychology*, **57**, 37–42.
Siegel, S., Siegel, A. E., and Andrews, J. M. (1964). *Choice, Strategy, and Utility*. New York: McGraw-Hill.
Simon, H. A. (1957). *Models of Man: Social and Rational*. New York: Wiley.
Simon, H. A. (1976). *Administrative Behavior: A Study of Decision-making Processes in Administrative Organizations*. 3rd ed. New York: Free Press.
Singleton, R. R. and Tyndall, W. F. (1974). *Games and Programs: Mathematics for Modeling*. San Francisco: W. H. Freeman.
Skotko, V., Langmeyer, D., and Lundgren, D. (1974). Sex differences as artifacts in the prisoner's dilemma game. *Journal of Conflict Resolution*, **18**, 707–13.
Slovic, P., Fischhoff, B., and Lichtenstein, S. (1977). Behavioral decision theory. *Annual Review of Psychology*, **28**, 1–39.
Smith, A. (1776). *An Inquiry Into the Nature and Causes of the Wealth of Nations*. Reprinted in an edition Ed. by E. Cannan, 1904. London: Methuen. Vol. I.
Smith, W. P. and Anderson, A. J. (1975). Threats, communication and bargaining. *Journal of Personality and Social Psychology*, **32**, 76–82.

Stahelski, A. J. and Kelley, H. H. (1969). The effects of incentives on cooperators and competitors in a Prisoner's Dilemma game. Paper presented at the Western Psychological Association Convention, Vancouver. [Cited in Wrightsman, O'Connor, & Baker (1972), pp. 51–2.]

Stavely, E. S. (1972). *Greek and Roman Voting and Elections.* Ithaca, N.Y.: Cornell University Press.

Steen, L. A. (1980). From counting votes to making votes count: The mathematics of elections. *Scientific American,* **243** (4), 16–26.

Stryker, S. (1972). Coalition behavior. In C. G. McClintock (Ed.), *Experimental Social Psychology,* pp. 338–80. New York: Holt, Rinehart & Winston.

Stryker, S. and Psathas, G. (1960). Research on coalitions in the triad: findings, problems, and strategy. *Sociometry,* **23**, 217–30.

Sullivan, M. P. and Thomas, W. (1972). Symbolic involvement as a correlate of escalation: the Vietnam case. In R. M. Russett (Ed.), *Peace, War, and Numbers.* Beverly Hills, Calif.: Sage.

Suppes, P. and Atkinson, R. C. (1960). *Markov Learning Models for Multiperson Interactions.* Stanford: Stanford University Press.

Swingle, P. G. (1970a). Dangerous games. In P. G. Swingle (Ed.), *The Structure of Conflict,* pp. 235–76. New York: Academic Press.

Swingle, P. G. (Ed.) (1970b). *The Structure of Conflict.* New York: Academic Press.

Tedeschi, J. T., Bonoma, T. V., and Lindskold, S. (1970). Threateners' reactions to prior announcements of behavioral compliance or defiance. *Behavioral Science,* **15**, 131–9.

Tedeschi, J. T., Bonoma, T. V., and Novinson, N. (1970). Behavior of a threatener: Retaliation vs. fixed opportunity costs. *Journal of Conflict Resolution,* **14**, 69–76.

Tedeschi, J. T., Schlenker, B. R., and Bonoma, T. V. (1973). *Conflict, Power, and Games.* Chicago: Aldine.

Teger, A. I. (1980). *Too Much Invested to Quit.* New York: Pergamon.

Teger, A. I. and Carey, M. S. (1980). Stages of escalation. In A. I. Teger, *Too Much Invested to Quit,* pp. 45–60. New York: Pergamon.

Teger, A. I., Carey, M. S., Hillis, J., and Katcher, A. (1980). Physiological and personality correlates of escalation. In A. I. Teger, *Too Much Invested to Quit,* pp. 61–81. New York: Pergamon.

Thibaut, J. W. and Kelley, H. H. (1959). *The Social Psychology of Groups.* New York: Wiley.

Thomson, J. J. (1979). Common-sense morality. *Journal of Philosophy,* **76**, 545–7.

Thorndike, R. L. (1911). *Animal Intelligence: Experimental Studies.* New York: Macmillan.

Tognoli, J. (1975). Reciprocation of generosity and knowledge of game termination in the decomposed Prisoner's Dilemma Game. *European Journal of Social Psychology,* **5**, 297–312.

Tropper, R. (1972). The consequences of investment in the process of conflict. *Journal of Conflict Resolution,* **16**, 97–8.

Trussler, S. (1973). *The Plays of Harold Pinter: An Assessment.* London: Victor Gollancz.

Tuck, R. (1979). Is there a free-rider problem, and if so, what is it? In R. Harrison (Ed.), *Rational Action: Studies in Philosophy and Social Science.* Cambridge: Cambridge University Press.

Tysoe, M. (1982). Bargaining and negotiation. In A. M. Colman (Ed.), *Cooperation and Competition in Humans and Animals,* pp. 141–72. Wokingham: Van Nostrand.

Tyszka, T. and Grzelak, J. L. (1976). Criteria of choice in non-constant zero-sum games. *Journal of Conflict Resolution,* **20**, 357–76.

Vinacke, W. E. (1969). Variables in experimental games: toward a field theory. *Psychological Bulletin,* **71**, 293–318.

Vinacke, W. E. and Arkoff, A. (1957). An experimental study of coalitions in the triad. *American Sociological Review,* **22**, 406–14.

von Neumann, J. (1928). Zur Theorie der Gesellschaftsspiele. *Mathematische Annalen,* **100**, 295–320.

von Neumann, J. and Morgenstern, O. (1944). *Theory of Games and Economic Behavior.* Princeton, N.J.: Princeton University Press. 2nd ed., 1947; 3rd ed., 1953.

Vorob'ev, N. N. (1977). *Game Theory: Lectures for Economists and Systems Scientists*. New York: Springer-Verlag. [Original Russian ed., *Teoriya Igr*, 1974, Leningrad University Press.]

Wald, A. (1945). Statistical decision functions which minimize maximum risk. *Annals of Mathematics*, **46**, 265–80.

Walker, M. (1977). The Jamaican fishing study reinterpreted. *Theory and Decision*, **8**, 265–72.

Weber, M. (1921). *Wirtschaft und Gesellschaft*. Tübingen: J. C. B. Mohr.

Wichman, H. (1972). Effects of isolation and communication on cooperation in a two-person game. In L. S. Wrightsman, J. O'Connor, and N. J. Baker (Eds), *Cooperation and Competition: Readings on Mixed-Motive Games*, pp. 197–205. Belmont: Brooks-Cole.

Wilke, H. and Mulder, M. (1971). Coalition formation on the gameboard. *European Journal of Social Psychology*, **1**, 339–56.

Williams, B. (1979). Internal and external reasons. In R. Harrison (Ed.), *Rational Action: Studies in Philosophy and Social Science*. Cambridge: Cambridge University Press.

Williams, J. D. (1966). *The Compleat Strategyst*, 2nd ed. New York: McGraw-Hill.

Willis, R. H. (1962). Coalitions in the tetrad. *Sociometry*, **25**, 358–76.

Wilson, R. (1975). On the theory of aggregation. *Journal of Economic Theory*, **10**, 89–99.

Wolff, R. P. (1968). *Kant: A Collection of Critical Essays*. London: Macmillan.

Worchel, S., Andreoli, V. A., and Folger, R. (1977). Intergroup cooperation and intergroup attraction: The effect of previous interaction and outcome of combined effort. *Journal of Experimental Social Psychology*, **13**, 131–40.

Wrightsman, L. S. (1966). Personality and attitudinal correlates of trusting and trustworthy behaviors in a two-person game. *Journal of Personality and Social Psychology*, **4**, 328–32.

Wrightsman, L. S., O'Connor, J., and Baker, N. J. (Eds) (1972). *Cooperation and Competition: Readings in Mixed-Motive Games*. Belmont: Brookes-Cole.

Young, J. W. (1977). Behavioral and perceptual differences between structurally equivalent, two-person games: A rich versus poor context comparison. *Journal of Conflict Resolution*, **21**, 299–322.

Zeckhauser, R. (1973). Voting systems, honest preferences, and Pareto optimality. *American Political Science Review*, **67**, 934–46.

Zermelo, E. (1912). Über eine Anwendung der Mengenlehre auf die Theorie des Schachspiels. *Proceedings of the Fifth International Congress of Mathematicians, Cambridge*, **2**, 501–10.

Index